Estimation of Regression Functions with certain Monotonicity and Concavity/Convexity Restrictions.

ULLA DAHLBOM

Göteborg University
GÖTEBORG
1994

Almqvist & Wiksell International
ISBN 91-22-01641-4

Printed in Sweden by
KOMPENDIET, Göteborg
1994

SUMMARY

In all regression problems the choice of estimation method is due to a priori information about the regression function. In many applications the relationship between dependent and independent variables is determined by some constraints on the variables. One of the simplest characteristics is monotonicity. The basic theory of isotonic regression is described in Barlow et al (1972) and Robertson et al (1989). A related problem is when the regression function consists of two phases - one non-decreasing and one non-increasing - with a known or unknown mode. In this situation we can use unimodal regression. This estimation method is contained in Frisén (1986). Further statistical properties are given in Dahlbom (1983, 1986). Another simple shape characteristic is concavity (or convexity). Serveral algorithms have been developed for concave regression. Some of the latest are those proposed by Wu (1982), Holm and Frisén (1985) and Fraser and Massam (1987).
In some applications it is warranted to consider regression functions with more specific similar characteristics. For instance, in quantal response assays in biological applications, and in modelling and forecasting human populations, sigmoid curves are used. Finney (1978) and Stone (1979) are suggesting a number of parametric sigmoid functions for analysis of such applications. Sigmoid curves are increasing functions, first increasing and convex up to some (usually unknown) inflection point and then increasing and concave.

In this report we propose a new estimation method that can be used for increasing and concave, increasing and convex, decreasing and concave and decreasing and convex functions, or suitable combinations of these four curves. The least squares estimate of such a regression function is a piecewise linear function with bending points contained in the set of the observed values of the independent variable. The set of bending points, which makes the function a least squares solution, can be determined by an iterative algorithm within a finite number of steps. If we have a sigmoid curve then two neighbouring bending points are determining the shift from the convex to the concave phase. A proof is given for the convergence of the algorithm.

We have restricted ourselves to look at some bias, variance and distribution properties of estimates for a few regression functions. In the first place the properties have been examined by calculations. In certain situations we have not managed to find the theoretical solutions. Then simulations have been used as a complement. Most simulations have been made for normally distributed y-variables obtained for uniformly distributed and equidistant x-variables. The results from many simulations are generalized to other similar situations.

In each linear piece of the LS solution we have positive bias for the estimated y-variables at both ends of the x-interval between two successive bending points and negative bias in the middle. For the whole regression function the average bias equals to zero. In symmetric situations we obtain unbiased estimates of x_{max} in unimodal regression and of the inflection point in sigmoid regression. Generally bias is affected by three factors; the curvature and the skewness of the regression function, the fact that the position of the x-values is randomized on the x-axis and the estimation method. For concave-unimodal regression the influences from these factors on bias of the estimated y_{max} are theoretically approximated.

Simulations show that the variances of the estimates of Y compared to Var(Y) are almost constant for all x-values in the middle of the regression curve. However in the end points these fractions increase. This means that the estimates suffer from border effects from the estimation procedure. The problem of estimating Var(y) when there is only one y-observation for each x is discussed. In this situation an estimation method is proposed using the assumption that we have the same variance for all y-observations.

As a complement to bias and variance properties it is often interesting to get some idea of the shape of the distribution of the examined estimator. We examine the distributions mostly by simulations where the 3:rd, μ_3, and the 4:th, μ_4, moments are estimated. For regression functions with constant curvature we also try to estimate a limit distribution of $\hat{Y}(X=x_i)$ using theoretical results concerning concave/convex regression obtained by Mammen (1988).

For distributions that belong to the exponential family it is sometimes possible to show that concave/convex regression estimates are in fact maximum likelihood estimates. This kind of generalization is discussed for the binomial distribution in connection with a biological problem.

We also suggest how to use the knowledge of bias, variance and the distribution to construct approximate confidence intervals.

Finally we have compared our estimation method to other similar methods. The conclusion from this comparison is that in many situations the proposed estimation method is a better choice.

CONTENTS:

1. INTRODUCTION.

In all regression problems the choice of estimation method is due to the a priori information about the regression function. In some applications the shape of the regression function is known and the estimation problem is reduced to an estimation of some parameters. But usually this is not the case. Instead the relationship between the dependent and independent variables is determined by some constraints on the variables. The simplest characteristic is monotonicity. In this case the only assumption about the regression function is, that it has a non-decreasing phase or/and a non-increasing phase. If the regression function consists of one non-decreasing (or non-increasing) phase only a suitable estimation method is isotonic regression. This is a non-parametric regression method that is very often referred to. The basic theory of isotonic regression is described in Barlow et al (1972) and Robertson et al (1989). If on the other hand the regression function consists of two phases - one non-decreasing and one non-increasing - with a known or unknown mode we can use unimodal regression. This estimation method is descibed in Frisén (1980,1986). Further statistical properties are given in Dahlbom (1983,1986).

Another simple shape characteristic is concavity (or convexity). In many applications it is motivated to consider regression functions with both monotonicity and concavity/convexity restrictions. The concave regression problem was first formulated by Hildreth (1954) for the estimation of marginal productivity curves. He adopted the LS estimation method and formulated it as a quadratic programming problem. Dent (1973) and Holloway (1979) continued the work and the concave regression function was obtained by a more general framework of quadratic programming with linear inequality constraints. The consistency of these LS concave regression estimators has been proved by Hanson and Pledger (1976). Wu (1982) proposed two algorithms for concave regression. One involves some quadratic programming. Unfortunately this algorithm giving the LS solution has no assured convergence. The other is just an approximation, but very easy to implement. An iterative method for concave regression was proposed by Holm and Frisén (1985). This method gives the LS estimate in a finite number of steps. Fraser and Massam (1987) formulated an algorithm using cylinder projections. Theoretically they obtained the LS solution but the estimation method involves a matrix inversion which in practice implies numerical uncertainty in the obtained results.

In many applications it is motivated to consider regression functions with more specific characteristics. In quantal response assays in biological applications for instance sigmoid curves are used. Finney (1978) has suggested a number of parametric sigmoid curves for analysis of such applications. Another area where sigmoid functions are widely used is in modelling and forecasting human populations. Stone (1979) eloquently details and describes these models. The two sigmoids that have received greatest attention are the logistic and the Gompertz models founded by Verhulst (1838) and Gompertz (1825) respectively. Many phenomena are either discrete to their nature, like plant reproduction or investment review, or can only be observed in a discrete fashion. In spite of this almost all estimation methods contain continuous models. Very few attempts have been made to formulate alternative regression procedures for sigmoids. A discrete alternative to the logistic model though was proposed by Nash (1977).

In this report we propose a new algorithm for all kinds of regression problems with different concavity/convexity restrictions, such as increasing and concave, unimodal and sigmoid regression. It is an iterative procedure for determining least squares estimates. The proposed estimation method is closely related to the estimation method proposed by Holm and Frisén (1985). The iteration with an inclusion and an exclusion procedure is identical for both estimation methods. In Holm and Frisén the estimation problem is solved by using a function system including a matrix inversion. The parameter estimates therefore may contain some numerical uncertainty. To avoid this problem our proposed method contains restrictions on the differences between two successive regression coefficients. Using these differences it is also easier to extend the problem from increasing and concave regression to unimodal and sigmoid regression. These two later problems are mentioned but not studied by Holm and Frisén.

In section 2 our proposed algorithm is described in different situations together with some proofs of general properties which are used in the estimation procedure. In chapter 3 the different simulation studies are described. We also discuss the generalization of the simulation results. A collection of theoretical properties is given in section 4 including the convergence of the algorithm. Chapters 5, 6 and 7 deal both with theory and simulations concerning bias, variances and distributions in some situations. A discussion on how to construct confidence intervals is given in chapter 8. In chapter 9 a biological problem is discussed and we propose a solution using ML estimation. The proposed solution is compared to an algorithm given in Schmoyer (1984). An example given by Schmoyer is also discussed and the two different estimation algorithms are compared. Finally in chapter 10 we compare the results from the proposed algorithm with other similar estimation procedures for increasing and concave, unimodal and sigmoid regression. Two examples, one given in Fraser and Massam (1987) and one in Wu (1982) concerning concave regression, and two given in Nash (1977) concerning sigmoid regression, are also discussed.

2. THE ESTIMATION PROCEDURE.

2.1. Solutions to some problems.

The purpose of this paper is to find a regression function with simple curve characteristics like isotonic restrictions and convexity/concavity. Thus for the observations (x_1, y_1), (x_2, y_2), ...(x_N, y_N) with $x_1 < x_2 .... < x_N$ and with nonnegative weights w_1, $w_2, ..., w_N$ we will search for a regression function $f(x_i)$ that minimizes

$$\sum_{i=1}^{N}[y_i - f(x_i)]^2 w_i$$

Before the estimation method is described we start by giving the solutions to some problems that will arise in the procedure. This will make the estimation procedure more easy to understand.

Problem 1:

Denote the observations by (x_1, y_1), (x_2, y_2),...,(x_N, y_N) and the corresponding non-negative weights by $w_1,...,w_N$. Determine the least squares estimate, $\hat{y}_i = y_0 + \hat{\beta}_0(x_i - x_0)$, of a regression line through a given point (x_0, y_0). We now minimize the sum of squares

$$Q_1 = \sum_{i=1}^{N}(y_i - y_0 - \hat{\beta}_0(x_i - x_0))^2 w_i$$

with respect to $\hat{\beta}_0$. The least squares estimate of β_0 becomes

$$\hat{\beta}_0 = \frac{\sum_{i=1}^{N}(y_i - y_0)(x_i - x_0) w_i}{\sum_{i=1}^{N}(x_i - x_0)^2 w_i}$$

The corresponding minimal sum of squares equals

$$\sum_{i=1}^{N}(y_i - y_0)^2 w_i - \frac{\left(\sum_{i=1}^{N}(y_i - y_0)(x_i - x_0) w_i\right)^2}{\sum_{i=1}^{N}(x_i - x_0)^2 w_i}$$

Problem 2:

The sum of squares Q_1 can be written as a sum of a quadratic expression of $(\bar{y} - y_0)$ and a part of the ordinary sum of squares determined by the least squares estimate of the regression line, where $\bar{y}$ is the weighted average for $y_i \neq y_0$. The sum of squared errors in problem 1 can be rewritten as

$$Q_1 = \sum_{i=1}^{N}[y_i - y_0 - \hat{\beta}_0(x_i - x_0)]^2 w_i =$$

$$[\bar{y} - y_0 - \hat{\beta}(\bar{x} - x_0)]^2 \sum_{i=1}^{N} w_i \, [1 - \frac{(\bar{x} - x_0)^2 \sum_{i=1}^{N} w_i}{\sum_{i=1}^{N} (x_i - x_0)^2 w_i}] + \sum_{i=1}^{N} [y_i - \bar{y} - \hat{\beta}(x_i - \bar{x})]^2 w_i .$$

where $\bar{x}$ is the weighted average for $x_i \neq x_0$. We denote

$$w_0 = \sum_{i=1}^{N} w_i \, (1 - \frac{(\bar{x} - x_0)^2 \sum_{i=1}^{N} w_i}{\sum_{i=1}^{N} (x_i - x_0)^2 w_i}) = \frac{\sum_{i=1}^{N} w_i \sum_{i=1}^{N} (x_i - \bar{x})^2 w_i}{\sum_{i=1}^{N} (x_i - x_0)^2 w_i}$$

We continue by looking at the sum of squares of errors. We have already shown that this sum is a quadratic function of $(\bar{y} - y_0)$ and that the minimal sum of squares depends on y_0 for a fixed x_0. Now we will determine which y_0-value that minimizes the sum of squares. This is done by derivation:

$$\frac{dQ_1}{dy_0} = 2(\bar{y}-y_0) \sum_{i=1}^{N} w_i \left[1 - \frac{(\bar{x}-x_0)^2 \sum_{i=1}^{N} w_i}{\sum_{i=1}^{N} (x_i-x_0)^2 w_i}\right] - \frac{2(\bar{x}-x_0) \sum_{i=1}^{N} w_i}{\sum_{i=1}^{N} (x_i-x_0)^2 w_i} \left[\sum_{i=1}^{N} (x_i-\bar{x})(y_i-\bar{y}) w_i\right].$$

The minimum is found by letting the derivate equal zero. Denote by y_0^* the value for which minimum occurs. Then

$$(\bar{y} - y_0^*) \sum_{i=1}^{N} w_i \left[1 - \frac{(\bar{x} - x_0)^2 \sum_{i=1}^{N} w_i}{\sum_{i=1}^{N} (x_i - x_0)^2 w_i}\right] = \frac{\sum_{i=1}^{N} (x_i - \bar{x})(y_i - \bar{y}) w_i}{\sum_{i=1}^{N} (x_i - x_0)^2 w_i} \left[\bar{x} - x_0\right] \sum_{i=1}^{N} w_i \quad \Leftrightarrow$$

$$(\bar{y} - y_0^{*}) = \frac{\sum_{i=1}^{N}(x_i - \bar{x})(y_i - \bar{y})w_i}{\sum_{i=1}^{N}(x_i - \bar{x})^2 w_i}(\bar{x} - x_0) = \hat{\beta}(\bar{x} - x_0)$$

where $\hat{\beta}$ is the regression coefficient in the ordinary weighted least squares estimate of the regression line. Thus the sum of squares of errors is minimized when $y_0^{*} = \bar{y} - \hat{\beta}(\bar{x} - x_0)$ ie when the predetermined point (x_0, y_0) lies on the ordinary least squares estimate of the regression line.

Problem 3:

The next problem deals with how big the loss will be in the sum of squares when we use linear regression through the predetermined point (x_0, y_0) instead of the ordinary weighted least squares estimate. The y_0-value that minimizes the sum of squares is still denoted y_0^{*}. Let Q_1 denote the sum of squares when using linear regression through the predetermined point (x_0, y_0) and Q_2 the sum of squares using the ordinary weighted least squares estimate, ie

$$Q_1 = \sum_{i=1}^{N}(y_i - y_0 - \hat{\beta}_0(x_i - x_0))^2 w_i \quad \text{and} \quad Q_2 = \sum_{i=1}^{N}(y_i - \bar{y} - \hat{\beta}(x_i - \bar{x}))^2 w_i.$$

THEOREM 2.1: The loss in the sum of squares is $(y_0 - y_0^{*})^2 w_0$, where

$$w_0 = \frac{\sum_{i=1}^{N} w_i \sum_{i=1}^{N} (x_i - \bar{x})^2 w_i}{\sum_{i=1}^{N} (x_i - x_0)^2 w_i},$$

when using linear regression through the predetermined point (x_0, y_0) instead of the ordinary weighted least squares estimate.

PROOF: The difference between Q_1 and Q_2 is

$$Q_1 - Q_2 = \sum_{i=1}^{N} (y_i - y_0 - \hat{\beta}_0 (x_i - x_0))^2 w_i - \sum_{i=1}^{N} (y_i - \bar{y} - \hat{\beta} (x_i - \bar{x}))^2 w_i =$$

$$\frac{\sum_{i=1}^{N} w_i \sum_{i=1}^{N} (x_i - \bar{x})^2 w_i}{\sum_{i=1}^{N} (x_i - x_0)^2 w_i} (y_0 - y_0^*)^2 = (y_0 - y_0^*)^2 w_0.$$

Thus the loss in the sum of squares is the weighted squared difference between the y-value we use for the predetermined point (x_0, y_0) and the y-value, y_0^*, for the same x-observation that minimizes the sum of squares.

Q.E.D.

2.2. Monotonic increasing and concave regression.

Suppose we have a set of observations (x_i, y_i) with weights w_i, i=1, ...,N. First we check if the observations themselves fulfill the monotonic increasing and concave restrictions. If this is the case, $f(x_i) = y_i$, i=1,...,N, is the solution to the problem and

$$\min \sum_{i=1}^{N} [y_i - f(x_i)]^2 w_i = 0.$$

But usually this is not the case and instead we have to use a more complicated estimation procedure. When a regression function is monotonic increasing and concave two conditions are fulfilled, namely

1) $\hat{y}_i < \hat{y}_{i+1}$ for $i=1,\ldots,(N-1)$

2) $\dfrac{\hat{y}_{i+1} - \hat{y}_i}{x_{i+1} - x_i} \geq \dfrac{\hat{y}_{i+2} - \hat{y}_{i+1}}{x_{i+2} - x_{i+1}}$ for $i=1,\ldots,(N-2)$

Equality in the 2:nd condition means that two or more estimated observations will lie on a straight line. This means that the regression function consists of continuous linear pieces where the estimate of the regression coefficient for each linear part will change in the first x-value of the corresponding regression line. Therefore the regression function will bend in these x-values and for this reason they will be called 'bending points'.

A major problem in the estimation method is to find the bending points that minimizes the sum of squares. We start the proposed estimation procedure by introducing some notations:

Suppose that the observations are divided into M ordered groups of some sizes, where $x_1 \leq \ldots \leq x_{i_1}$ belong to the first group, $x_{i_1+1} \leq \ldots \leq x_{i_2}$ to the second and so on.

DEF: The smallest x-observation in each group is called the bending point for the corresponding x-interval.

Since bending points will appear a few times in the description of the estimation method they will have a special notation, $x_0^{(k)}$, where "k" is the interval in which the bending point lies, ie $x_0^{(1)}=x_1$, $x_0^{(2)}=x_{i_1+1}$, $x_0^{(3)}=x_{i_2+1}$ and so on.

We will also need a notation for a predicted y-value corresponding to the bending point. Therefore we introduce the notation and the definition of this predicted value according to the following:

Let $x_0^{(k)}$ be the bending point of the k:th x-interval. Predict a y-value, $y_0^{(k)}$, for $x_0^{(k)}$ using the weighted least squares estimate of the regression line in the (k-1):th x-interval. The estimate is obtained from the observed y-values in this x-interval together with a predicted y-value, corresponding to the bending point $x_0^{(k-1)}$, with the weight $w_0^{(k-1)}$, obtained from the regression line in the (k-2):th x-interval.

DEF. A predicted point, $(x_0^{(k)}, y_0^{(k)})$, where $x_0^{(k)}$ is the bending point of the k:th x-interval and $y_0^{(k)}$ is the corresponding y-value, predicted according to the above description, is called the fiction point from the k:th x-interval.

Since it contains a few different parts, we first demonstrate how the procedure works on fixed bending points. This case could result in a solution that does not fulfill the isotonic and concave restrictions.

2.2.1. Fixed bending points.

We can distinguish two steps in the estimation procedure. The only purpose of the first step is to predict the fiction point, $(x_0^{(k)}, y_0^{(k)})$ and the corresponding weight for each x-interval. This weight contains all information of the sum of squares from the preceding x-intervals. These fiction points with corresponding weights are used in the second step of the estimation procedure to determine the least squares estimate of the continuous piecewise linear regression function.

Let $\bar{y}_k$ and $\bar{x}_k$ denote the weighted averages of the y- and x-observations in the k:th x-interval and denote the ordinary weighted least squares estimate of β in the same interval by b_k. Distinction is made between b_k and $\hat{\beta}_k$, which is the notation of the weighted least squares estimate of β_k of the straight line trough a predetermined point. However, the use of the notation $\hat{\beta}_k$ will not be of current interest until we come to the second step in the estimation procedure concerning continuous piecewise linear regression. In a corresponding way distinction is made between $f_k(x)$ and $f_k^*(x)$. The notation $f_k(x)$ means the ordinary weighted least squares estimate of a regression line and $f_k^*(x)$ means the weighted least squares estimate of a regression line through a predetermined point.

STEP 1:

Determine the ordinary weighted least square estimate, $f_1(x) = \bar{y}_1 + b_1(x - \bar{x}_1)$, of the linear regression line in the first x-interval $x_1 \leq \ldots\ldots \leq x_{i_1}$. Use this straight line to predict a y-value corresponding to the bending point, $x_{i_1+1} = x_0^{(2)}$, in the second interval

$$f_1(x_{i_1+1}) = \bar{y}_1 + b_1(x_{i_1+1} - \bar{x}_1)$$

This fiction point $(x_0^{(2)}, y_0^{(2)}) = (x_{i_1+1}, f_1(x_{i_1+1}))$ will, according to the discussion in problem 2, have the weight

$$w_0^{(2)} = \frac{\sum_{k=1}^{i_1} w_k \sum_{k=1}^{i_1} (x_k - \bar{x}_1)^2 w_k}{\sum_{k=1}^{i_1} (x_k - x_0^{(2)})^2 w_k}$$

where $x_0^{(2)}$ is the bending point in the second x-interval. Thus, to the second x-interval, $x_{i_1+1} \leq \ldots\ldots\ldots \leq x_{i_2}$, we have added an extra observation, which includes all information about the sum of squares in the first x-interval.

Next determine the ordinary weighted least squares estimate, $f_2(x) = \bar{y}_2 + b_2(x - \bar{x}_2)$, of the regression line through the second x-interval containing the observed points $(x_{i_1+1}, y_{i_1+1}), \ldots\ldots\ldots\ldots, (x_{i_2}, y_{i_2})$ including the fiction point $(x_0^{(2)}, y_0^{(2)})$. Use $f_2(x)$ to predict a y-value for the bending point, $x_{i_2+1} = x_0^{(3)}$, in the third interval. This new fiction point $(x_0^{(3)}, y_0^{(3)}) = (x_{i_2+1}, f_2(x_{i_2+1}))$ will have the weight

$$w_0^{(3)} = \frac{\sum_{k=i_1+1}^{i_2} w_k \sum_{k=i_1+1}^{i_2} (x_k - \bar{x}_2)^2 w_k}{\sum_{k=i_1+1}^{i_2} (x_k - x_0^{(3)})^2 w_k}$$

where $\bar{x}_2$ is the weighted average of the ordinary x-observations in the second interval including the bending point $x_0^{(2)}$. Thus the third x-interval now contains an extra observation, which includes all information about the sum of squares in the two first x-intervals. This procedure will continue in the same way up to the last x-interval, where we beyond the observed points will have a fiction point with a calculated weight which contains all information of the sum of squares from all the former x-intervals. To this last x-interval we will determine the weighted least squares estimate of the regression line through the observations including the fiction point $(x_0^{(M)}, y_0^{(M)})$ as before. Since in this case we have restricted the problem to <u>increasing</u> and concave regression, we must check if $b_M \geq 0.0$:

1) If this condition is fulfilled we can continue the procedure in its second step.

2) If $b_M < 0.0$ then we estimate β_M by b_M=0.0 and $f_M(x)$ by $\bar{y}_M$, where $\bar{y}_M$ is the weighted average of the y-observations in the M:th x-interval including the y-value in the fiction point. After this adjustment we continue the procedure in its second step.

STEP 2:

Up to this point in the procedure we have estimated M straight lines $f_k(x)$, k = 1,...,M. The only purpose of this is to calculate the weights, $w_0^{(k)}$, and the fiction points $(x_0^{(k)}, y_0^{(k)})$, k=2,...,M. To determine the least squares estimate of the continuous piecewise linear regression we must continue the procedure in its second step. In doing so we will use the solutions to the problems mentioned in the beginning of this chapter.

In the piecewise procedure we start at the last x-interval. In this interval we have already estimated a straight line for the observed values and the fiction point $(x_0^{(M)}, y_0^{(M)})$.

This means that in the M:th x-interval we obtain $b_M = \hat{\beta}_M$ and $f_M(x) = f_M^*(x)$ and the estimate of the regression line can be denoted by $f_M^*(x) = \bar{y}_M + \hat{\beta}_M(x - \bar{x}_M)$. Calculate the value of $f_M^*(x)$ for the bending point, $x_0^{(M)}$, in the last interval. Next determine the least squares estimate of the regression line in the second last x-interval including the fiction observation, $(x_0^{(M-1)}, y_0^{(M-1)})$, in a way that will make the regression line go through the predicted point, $(x_0^{(M)}, f_M^*(x_0^{(M)}))$, in the last x-interval. Since the weight, $w_0^{(M-1)}$, of the fiction point $(x_0^{(M-1)}, y_0^{(M-1)})$ contains all information from the sum of squares of the first (M-2) x-intervals, this method will give us the minimum sum of squares in the (M-1):th x-interval taking the first (M-2) x-intervals into account.

According to the discussion in problem 1 earlier in this chapter the least squares estimate of β will be $\hat{\beta}_{M-1} = K_1/K_2$ where

$$K_1 = \sum_{i=1}^{m_{(M-1)}}[y_i-y_0^{(M)}][x_i-x_0^{(M)}]w_i + [f_{M-2}(x_0^{(M-1)})-y_0^{(M)}][x_0^{(M-1)}-x_0^{(M)}]w_0^{(M-1)}$$

and

$$K_2 = \sum_{i=1}^{m_{(M-1)}}[x_i - x_0^{(M)}]^2 w_i + [x_0^{(M-1)}-x_0^{(M)}]^2 w_0^{(M-1)}$$

The notation $m_{(M-1)}$ means the number of observations in the second last x-interval excluding the fiction point, $(x_0^{(M-1)}, y_0^{(M-1)})$, and the weighted sums symbolize the summary of these observations. We have now connected two straight lines, one in each of the two last x-intervals, and the connection point is $[x_0^{(M)}, f_M^*(x_0^{(M)})] = [(x_0^{(M)}, f_{M-1}^*(x_0^{(M)})]$.

Next calculate the value of the regression line $f_{M-1}^*(x)$ in the bending point, $x_0^{(M-1)}$, of the second last x-interval. Estimate the weighted least squares regression line through this point $[x_0^{(M-1)}, f_{M-1}^*(x_0^{(M-1)})]$ for the third last x-interval including the fiction point containing information from the first to the (M-3):th x-interval. Now the straight lines, one in each of the 2:nd last and the

3:rd last x-intervals, are also connected and the connection point is $[x_0^{(M-1)}, f_{M-1}^*(x_0^{(M-1)})] = [x_0^{(M-1)}, f_{M-2}^*(x_0^{(M-1)})]$.
This procedure is repeated until all x-intervals are connected with weighted least squares estimates of regression lines through successively calculated points, $[x_0^{(i)}, f_i^*(x_0^{(i)})]$ and we have a continuous curve. Since the weights $w_0^{(i)}$, i=1,...,M, contain all information from the sum of squares of the preceding (i-1) x-intervals the procedure will give us the least squares estimate of the whole curve under the increasing and concave restictions. This holds both for $\hat{\beta}_M > 0.0$ and $\hat{\beta}_M = 0.0$. We now have to check if the restrictions 'monotonic increasing' and 'concave' are fulfilled. This is done by checking if $\hat{\beta}_{k-1} > \hat{\beta}_k$ for k=2,....,M. As mentioned before this does not have to be the case when the bending points are fixed.

We have described the estimation method where "step 1" starts in the first x-interval and the fiction points with corresponding weights are calculated up to the last x-interval. We then continue with "step 2" by estimating the continuous piecewise linear regression function from the last x-interval to the first. This is in fact one case out of several alternatives, which give the same unique solution. In the addition to the above solution it is also possible to start the whole estimation procedure with "step 1" in the last x-interval and calculate fiction points with corresponding weights down to the first x-interval. We can continue with "step 2" using continuous piecewise linear regression from the first up to the last x-interval. We can then check if the restrictions, ie monotonic increasing and concave, are fulfilled.
The procedure can just as well start with "step 1" from both the first and the last x-intervals simultaneously until we get estimated

bending points in each end of an inner interval. We continue with "step 2" by determining a least squares regression function in this x-interval and successively use continuous piecewise linear regression in each direction, both to the first and the last x-intervals. It is obvious that especially the last method gives some understanding of how the estimation method works.
As stated before, in each of these three cases we get the same unique solution.

2.2.2. Changing bending points.

The whole procedure starts by letting every observed value serve as a bending point. That means that we check if the observations themselves fulfill the isotonic and concave restrictions. If this is not the case we start an estimation procedure that uses an inclusion-exclusion method, which is described in Holm and Frisén (1985) according to the following:
Determine the ordinary weighted least squares estimate of the regression line for all the observations. Thus we have only one bending point, namely x_1. To examine which other points that could be possible bending points we reformulate the problem. Let

$$f_1(x) = 1$$

$$f_k(x) = \begin{cases} x - x_k & \text{for } x < x_k \\ 0 & \text{for } x \geq x_k \end{cases} \qquad k = 2, 3, \ldots, N$$

We can also write any non-decreasing concave regression function f(x) as $f(x) = \sum_{k=1}^{N} a_k f_k(x)$, where a_1 has no restrictions and $a_k \geq 0$ for $k \geq 2$. Thus our weighted least squares estimate is obtained by minimizing

$$\sum_{j=1}^{N}(y_j - \sum_{k=1}^{N} a_k f_k(x))^2 w_j.$$

1) Inclusion-procedure:

Let $\{I_0\}$ be the set of index-values corresponding to the x-observations that are bending points. When we start the procedure $\{I_0\}$ only contains one value namely i=1. Denote $f(x) = \sum_{k=1}^{N} a_k f_k(x)$

A. Calculate the scalar product between $f_j(x)$ and $f_k(x)$ for every k $\notin \{I_0\}$. Divide the product by the norm of $f_j(x)$:

$$\rho_j = \frac{[f(x) - \sum_{k \notin \{I_0\}} a_k f_k(x),\ f_j(x)]}{[f_j(x), f_j(x)]^{1/2}}$$

A1. If $\rho_j \leq 0$ for every j then we have found the least squares solution.

A2. If $\rho_j > 0$ for some j $\notin \{I_0\}$ denote the largest positive ρ-value by ρ_{ji} and let the corresponding index, ji, be included in the set $\{I_0\}$. This index set now contains one further index value and we have also got one further bending point. Use the procedure described for fixed bending points, which corresponds to index values contained in $\{I_0\}$. Estimate $f_k(x) = \alpha_k + \beta_k x$ where $k \in \{I_0\}$. Compare $\hat{\beta}_{k-1}$ to $\hat{\beta}_k$.

i) If the monotonic increasing and concave restrictions are fulfilled, ie if $\hat{\beta}_{k-1} \geq \hat{\beta}_k$ for k-1, k $\in \{I_0\}$ then start again from point A.

ii) If $\hat{\beta}_{k-1} < \hat{\beta}_k$ for some k-1, k $\in \{I_0\}$ then start the exclusion-part.

2) Exclusion procedure:

Go back to the former solution, which was the last solution, where the monotonic increasing and concave restrictions were fulfilled. Denote the estimated β_k-values in this solution by $\hat{\beta}_k^{+}$. Let

$$\gamma_k = \hat{\beta}_{k-1}^{+} - \hat{\beta}_k^{+}$$

$$\lambda_k = \hat{\beta}_{k-1} - \hat{\beta}_k .$$

Calculate

$$\varepsilon_k = \frac{\gamma_k}{\gamma_k - \lambda_k}$$

for k $\in \{I_0$- the index for the last obtained bending point$\}$.

B. If there exsists a minimal ε-value < 1 then let $\varepsilon_{jk} = \min \varepsilon_k < 1$, where minimum is obtained for k $\in \{I_0\}$. The index value jk is excluded from the index set $\{I_0\}$ and x_{jk} will no longer be a bending point. Thus we have one bending point less compared to the last time we used the procedure for fixed bending points.

The minimum value ε_{jk} is used to obtain a new estimate of β_k, k=1,......,M, where the monotonic increasing and concave restrictions are fulfilled. This is obtained by calculating a new weighted estimate of β_k, denoted by $\hat{\beta}_k^{w}$, according to $\hat{\beta}_k^{w} = \varepsilon_{jk} * \hat{\beta}_k + (1 - \varepsilon_{jk}) * \hat{\beta}_k^{+}$ where $k \in \{I_0\}$.

Next we start the procedure for fixed bending points again using the bending points x_k, $k \in \{I_0\}$. Estimate $f_k(x) = \alpha_k + \beta_k x$ where $k \in \{I_0\}$. Compare these new estimates $\hat{\beta}_{k-1}$ to $\hat{\beta}_k$.

B1. If the isotonic and concave restrictions are fulfilled ie if $\hat{\beta}_{k-1} \geq \hat{\beta}_k$ for k-1, k ∈ $\{I_0\}$ then start the inclusion procedure again from point A.

B2. If $\hat{\beta}_{k-1} < \hat{\beta}_k$ for some k-1, k ∈ $\{I_0\}$ then continue with the exclusion part. Go back to the former solution, which was the last solution, $\hat{\beta}_k^{\,w}$, where the monotonic increasing and concave restrictions were fulfilled.

Let

$$\nu_k = \hat{\beta}_{k-1}^{\,w} - \hat{\beta}_k^{\,w}$$

$$\delta_k = \hat{\beta}_{k-1} - \hat{\beta}_k.$$

Then calculate a new set of ε_k:

$$\varepsilon_k = \frac{\nu_k}{\nu_k - \delta_k}$$

After that the exclusion procedure continues from point B.

This inclusion-exclusion part of the estimation method will go on until $\rho_k \leq 0$ for every k ∉ $\{I_0\}$ at point A1 in the inclusion procedure. We have then received the least squares estimate of the regression function.

2.3. General adjustments in the estimation method when changing the concave/convex and the up/down restrictions.

The estimation method is of course not limited to just monotonic increasing and concave regression. We can just as well use it for monotonic decreasing and concave regression, isotonic and convex regression, pure convex, pure concave and suitable combinations of regression functions of the shapes concave-up, concave-down, convex-up and convex-down. But for each of these four cases we must make some adjustments in the continuous piecewise linear regression, use different function systems in the inclusion procedure and also change the restrictions for ρ_k and the comparisons between $\hat{\beta}_{k-1}$ and $\hat{\beta}_k$ in the inclusion-exclusion procedure.

2.3.1. Changes in the estimation procedure:

Concave-up:

In the estimation procedure for concave-up regression we had two restrictions for the estimates of β_k. One condition concerned the estimate of β_M, the regression coefficient in the last x-interval, namely $\hat{\beta}_M \geq 0$. This restriction was taken care of according to the following:

The weights, $w_0^{(k)}$, of the fiction points were calculated up to the last x-interval, denoted M. In the last x-interval we estimated the weighted least squares regression line. We then checked if $\hat{\beta}_M \geq 0$ before we continued the procedure in its second step. If $\hat{\beta}_M < 0$ we estimated β_M by $\hat{\beta}_M = 0.0$ and the corresponding straight line by $f_M^*(x) = \bar{y}_M$, where $\bar{y}_M$ was the weighted average of the y-observations in the M:th x-interval including the y-value of the fiction point. After this adjustment we continued the procedure in its second step.

Convex-down:

For the case of convex-down regression the corresponding restriction $\hat{\beta}_M \leq 0$ must be fulfilled. Here we estimate the weights, $w_0^{(k)}$, of the fiction points down to the last x-interval, denoted M. In the last x-interval we calculate the weighted least squares estimate of the regression line. Then we check if $\hat{\beta}_M \leq 0$ before we continue the procedure in its second step.

If $\hat{\beta}_M > 0$ we estimate β_M by $\hat{\beta}_M = 0.0$ and the corresponding straight line by $f_M^*(x) = \bar{y}_M$, where $\bar{y}_M$ is the weighted average of the y-observations in the M:th x-interval including the y-value of the fiction point. After this adjustment we continue the procedure in its second step.

Concave-down:

When we have a concave-down regression we must adjust the corresponding restriction for $\hat{\beta}_1$ in the following way:

If in the second step in the estimation procedure we obtain $\hat{\beta}_1 < 0$ we must start the regression method all over again.

This second time we start by estimating a straight line in the first x-interval in order to calculate the weight $w_0^{(2)}$. Now we check the estimate of β_1, denoted b_1 in the first step. If $b_1 \geq 0.0$ we continue the procedure. If $b_1 < 0.0$ we estimate β_1 by $b_1 = 0.0$. This estimate gives us the weight $w_0^{(2)} = \sum_{k=1}^{i_1} w_k$ corresponding to the fiction point $[x_0^{(2)}, f_2(x_0^{(2)})]$.

We now continue the procedure until in its second step we obtain an estimate of β_1. If $\hat{\beta}_1 \geq 0$ then we have finished the estimation procedure and can continue to the inclusion-exclusion part. If $\hat{\beta}_1 < 0$ we estimate β_1 by $\hat{\beta}_1 = 0$. We get the level of the regression line in the first x-interval by looking at the continuous piecewise regression line in the second x-interval. In the second bending point, $x_0^{(2)}$, we have an estimate of the corresponding y-value. Denote this y-value by $\hat{y}_0^{(2)}$. The estimate of the continuous piecewise regression line in the first x-interval will be $\hat{y} = \hat{y}_0^{(2)}$. Now we can continue to the inclusion-exclusion part.

Convex-up:

One of the necessary restrictions for convex-up regression is $\hat{\beta}_1 \leq 0$. This is checked and taken care of in a way that corresponds to the adjustments for concave-down procedure.

2.3.2. The function systems in the inclusion-part:

The function system for each of these four cases is specific according to the following:

Concave-up:

Let

$$f_1(x) = 1$$

$$f_k(x) = \begin{cases} x - x_k & \text{for } x < x_k \\ 0 & \text{for } x \geq x_k \end{cases} \qquad k = 2, 3, \ldots, N$$

Then we can write any non-decreasing concave regression function f(x) as $f(x) = \sum_{k=1}^{N} a_k f_k(x)$, where a_1 has no restrictions and $a_k \geq 0$ for $k \geq 2$. For the other three cases the following holds

Concave-down:
Let

$$f_0(x) = 1$$

$$f_k(x) = \begin{cases} 0 & \text{for } x \le x_k \\ x_k - x & \text{for } x > x_k \end{cases} \qquad k= 1, 2,.....,N-1$$

Then we can write any non-decreasing concave regression function f(x) as $f(x) = \sum_{k=0}^{N-1} a_k f_k(x)$, where a_0 has no restrictions and $a_k \le 0$ for $k \ge 1$.

Convex-up:
For

$$f_0(x) = 1$$

$$f_k(x) = \begin{cases} 0 & \text{for } x \le x_k \\ x - x_k & \text{for } x > x_k \end{cases} \qquad k= 1, 2,....,N-1$$

we can write any non-decreasing convex regression function f(x) as $f(x) = \sum_{k=0}^{N-1} a_k f_k(x)$, where a_0 has no restrictions and $a_k \ge 0$ for $k \ge 1$.

Convex-down:
Let

$$f_1(x) = 1$$

$$f_k(x) = \begin{cases} x_k - x & \text{for } x < x_k \\ 0 & \text{for } x \ge x_k \end{cases} \qquad k= 2, 3,....,N$$

In that situation we can write any non-increasing and convex regression function f(x) as $f(x) = \sum_{k=1}^{N} a_k f_k(x)$, where a_1 has no restrictions and $a_k \le 0$ for $k \ge 2$.

2.3.3. The changing of ρ_j in the inclusion-part:

One of the restrictions in the inclusion part, that is changed for different curve characteristics, is ρ_j. For concave regression functions the inclusion-exclusion procedure continues until $\rho_j \leq 0$ for every $j \notin \{I_0\}$ and for convex regresison functions it continues until $\rho_j \geq 0$ for every $j \notin \{I_0\}$.

2.3.4. The changing of order relations between $\hat{\beta}_{j-1}$ and $\hat{\beta}_j$.

Another of the restrictions that is connected to the convexity/ concavity property is the comparison between $\hat{\beta}_{j-1}$ and $\hat{\beta}_j$. When in the same procedure we compare $\hat{\beta}_{j-1}$ and $\hat{\beta}_j$ we must have the following order relations fulfilled:

Concave-up: $\hat{\beta}_{j-1} > \hat{\beta}_j \geq 0$

Concave-down: $0 \geq \hat{\beta}_{j-1} > \hat{\beta}_j$

Convex-up: $0 \leq \hat{\beta}_{j-1} < \hat{\beta}_j$

Convex-down: $\hat{\beta}_{j-1} < \hat{\beta}_j \leq 0.$

2.3.5. General remarks.

We can also use the estimation method by combining two of these four regression functions. Then we get some special curves. For instance two functions of the shapes concave-up and concave-down will form a specific type of unimodal function (we will also get this type of unimodal function if we use concave regression without monotonicity restrictions), the convex-up shape combined with concave-up will form a sigmoid function. These two special regression functions will have many analogous properties. When in this report we consider sigmoid regression functions we could therefore often mean unimodal functions or other combinations of two regression functions with isotonic and convex/concave restrictions. However, unimodal regression is much simpler since we don´t have any inflection point as in sigmoid regression.

2.4. Sigmoid regression.

This section will deal with the problem of finding a least squares estimate of a sigmoid regression function. When we combine the functions, $f_1^*(x),....,f_k^*(x)$, with convex-up and $f_{k+1}^*(x),....,f_M^*(x)$, with concave-up restrictions we have the following order relations of the corresponding regression coefficients fulfilled

$$0 \leq \hat{\beta}_1 < \hat{\beta}_2 .... < \hat{\beta}_k$$

$$\hat{\beta}_{k+1} > ....> \hat{\beta}_M \geq 0.$$

It is obvious that comparing $\hat{\beta}_k$ to $\hat{\beta}_{k+1}$ we get one out of two order relations between them.

1) $\hat{\beta}_k > \hat{\beta}_{k+1}$. Thus $\hat{\beta}_k$ can be contained both in the convex-up and the concave-up part of the sigmoid function.

2) $\hat{\beta}_k < \hat{\beta}_{k+1}$. Thus $\hat{\beta}_{k+1}$ can be contained both in the convex-up and the concave-up part of the sigmoid function.

From these two cases we see that the estimated sigmoid regression function has an inflection interval. Theorem 4:3 shows that the estimated inflection interval is always obtained between two neighbouring x-observations. This application of the estimation method will include the new problem of finding the inflection interval between two functions with isotonic and convex/concave restrictions.

In the estimation method the combination of the convex-up and concave-up restrictions will result in a special case for the function system in the inclusion-part according to the following. Let the index-value of the inflection interval be denoted by jk. Then the function system will be:

If $k \leq jk$ then

$$f_k(x) = \begin{cases} 0 & \text{for } x \leq x_k \\ x - x_k & \text{for } x > x_k \end{cases} \qquad k = 1, 2, \ldots, N-1$$

If $k > jk$ then

$$f_k(x) = \begin{cases} x - x_k & \text{for } x < x_k \\ 0 & \text{for } x \geq x_k \end{cases} \qquad k = 2, 3, \ldots, N$$

If in the estimation method we obtain $\hat{\beta}_1 < 0.0$ and/or $\hat{\beta}_M < 0.0$ then, as described before, we must estimate β_1 by $\hat{\beta}_1 = 0.0$ and/or β_M by $\hat{\beta}_M = 0.0$. These plateaus will affect the function system in the following way:

<u>For $k \leq jk$</u>

If $\hat{\beta}_M > 0.0$ then

$$f_k(x) = \begin{cases} 0 & \text{for } x \leq x_k \\ x - x_k & \text{for } x_k < x \end{cases} \qquad k = 1, 2, \ldots, N-1$$

If $\hat{\beta}_M = 0.0$ then

$$f_k(x) = \begin{cases} 0 & \text{for } x \le x_k \\ x - x_k & \text{for } x_k < x < x_0^{(M)} \\ x_0^{(M)} - x_k & \text{for } x_0^{(M)} \le x \end{cases} \qquad k = 1, 2, \ldots, N-1$$

<u>For k > jk</u>

If $\hat{\beta}_1 > 0.0$ then

$$f_k(x) = \begin{cases} x - x_k & \text{for } x < x_k \\ 0 & \text{for } x \ge x_k \end{cases} \qquad k = 2, 3, \ldots, N$$

If $\hat{\beta}_1 = 0.0$ then

$$f_k(x) = \begin{cases} x_0^{(2)} - x_k & \text{for } x \le x_0^{(2)} \\ x - x_k & \text{for } x_0^{(2)} < x < x_k \\ 0 & \text{for } x \ge x_k \end{cases} \qquad k = 2, 3, \ldots, N$$

If the partition in one convex-up and one concave-up phase is known, then the least squares estimate of f(x) will be obtained by using monotonic increasing and convex regression on the first phase and monotonic increasing and concave regression on the other phase. The inflection interval will be the x-interval that connects the two regression functions. But usually the partition is unknown. In this situation we can use the following procedure:

Suppose that we have the observations $(x_1, y_1), (x_2, y_2), \ldots\ldots (x_N, y_N)$ with $x_1 < x_2 \ldots\ldots < x_N$ and with non-negative weights $w_1, w_2, \ldots, w_N$. First we check if the inflection interval is to the left of the smallest observed x-value, x_1. Thus we have just the concave-up part of the sigmoid regression function. Calculate the sum Q_1 of weighted squared deviations between the observed and the estimated values of y_i.

Now suppose that the partition in phases is between x_1 and x_2. Since x_1 is just a single observation we get the estimate $\hat{y}_1 = y_1$. On the other part $x_2 \leq x \leq x_N$ we use monotonic increasing and concave regression. Calculate the sum Q_2 of weighted squared deviations between the observed and the estimated values of y_i.

This procedure is repeated for every partition until we come to the next special case where the inflection interval is to the right of the biggest observed x-value and we only get a monotonic increasing and convex regression function.

The corresponding sums Q_i, i=3,..,N, of weighted squared deviations between the observed and the estimated values of y_i are calculated. Thus for each partition we have estimated a sigmoid regression function and calculated the corresponding sum of squares of errors. For each partition the procedure gives us the least squares estimate among all combinations of fixed bending points. To get the solution we must compare all sums Q_i. The smallest of those Q_i-values corresponds to the weighted least squares estimate of the regression function and the estimate of the inflection interval is the estimated regression line that connects the two phases convex-up and concave-up. The estimate of the inflection point of the sigmoid curve is the midpoint of the estimated inflection interval.

This estimation procedure may well be done by computer and a FORTRAN program is available.

3. SIMULATIONS.

We think that the different properties essentially depend on the second order derivative. But we have not managed to develop a general theory. Instead we have used simulations. We have restricted ourselves to some simple situations but the obtained results are possible to generalize to other similar cases by transformations.

The simulation studies are made for normally, and in some cases, for binomially distributed variables. The corresponding regression functions fulfill restrictions on either isotonic concavity/convexity, unimodal concavity/convexity or sigmoidity. Examples of such regression functions $\mu_i(x)$ are

in the cases of isotonicity or unimodality:

1) $\mu_i(x) = a - cx^2$ $\qquad c > 0$

2) $\mu_i(x) = a - cx^2 + dx^3$ $\qquad 0.8 \le c \le 3.0, \quad 0.00 \le d \le 0.03$

3) $$\mu_i(x) = a \begin{cases} cx \text{ if } x \le 0 \\ -dx \text{ if } x > 0 \end{cases} \qquad c > 0,\ d > 0$$

and in cases of sigmoidity:

4) $$\mu_i(x) = a + 10 * \frac{e^x}{1 + e^x}$$

5) $$\mu_i(x) = a + 10* \int_{-\infty}^{x} \phi(t)\, dt = a + 10*\Phi(x)$$

6) $\mu_i(x) = a + \begin{cases} 0.0 & \text{if } x \le -2.5 \\ 2x + 5 & \text{if } -2.5 < x < 2.5 \\ 10.0 & \text{if } x \ge 2.5 \end{cases}$

7) $\mu_i(x) = a + 10*(1-e^{-e^x})$

Simulations using binomially distributed variables have only been performed for a few sigmoid curves like the logit, probit and extreme value functions. The position of x_{max} or x_{inf} in relation to the observed x-values is randomized. This is done because in reality we usually don't know the location of x_{max} or x_{inf}.
In the simulations we have used boundaries [a,b] of x-observations. These boundaries are chosen to be symmetric around x_{max} in cases of unimodality and around x_{inf} in cases of sigmoidity. The procedure will give us the estimate of the regression function as a continuous function consisting of piecewise straight lines. In chapter 4 we have shown that the influence of the observations in the (i-k):th or (i+k):th x-interval on the weighted estimated straight line in the i:th x-interval is at most $(0.275)^k$. If $k > 4$ this influence can probably be supposed to be negligible. This means that if we have several x-intervals and we want to estimate y_{max} or y_{inf}, the influence of observations in an interval a few steps away will not have an essential influence on the estimated maximum point or inflexion point. Suppose we have an isotonic or unimodal regression function of the type described before. Let c be the notation of the constant used in the chosen model. Then we can use $-5/\sqrt{c} \le x \le 5/\sqrt{c}$ as a reasonable interval [a,b]. In cases of sigmoidity the corresponding examined interval is $-5 \le x \le 5$. In most simulations 25 000 replications have been used.

Suppose we have the observations $Y_1,\ldots, Y_N$, corresponding to $x_1,\ldots, x_N$. Then for $x_i < x_k < x_l$ the procedure for concave regression will give us the estimates $\hat{Y}(x_i)$, $\hat{Y}(x_k)$ and $\hat{Y}(x_l)$ where according to concavity

$$\hat{Y}(x_l) \leq \frac{\hat{Y}(x_k)[x_l-x_i] + \hat{Y}(x_i)[x_k-x_l]}{x_k - x_i}.$$

According to lemma 4:2 the set of concave functions is a convex cone. This can be used to prove the following theorem concerning scale- and translation properties.

THEOREM 3.1:

1) If an observation set is translated k units in the x direction the concave isotonic estimate is also translated k units in the x direction,

2) if an observation set is translated k units in the y direction the concave isotonic estimate is also translated k units in the y direction,

3) if an observation set is scaled with a factor c in the x direction the concave isotonic estimate is also scaled with a factor c in the x direction,

4) if an observation set is scaled with a factor c in the y direction the concave isotonic estimate is also scaled with a factor c in the y direction.

PROOF: The four assertions follow directly from the relations between $\hat{Y}(x_i)$, $\hat{Y}(x_k)$ and $\hat{Y}(x_l)$ and the fact that the set of concave functions is a convex cone.

Q.E.D

COMMENT: The theorem also means that properties of the concave isotonic estimates can be transformed from one case to all cases obtained by the four types of translation and scale changes in the theorem.

Since in this report we mostly use observations from the normal distribution, the generalization in these situations can be done by rescaling according to the following:

Suppose we have a simulation study with the observations (x_i, y_{ij}), $j=1,\ldots,n_1$, $i=1,\ldots,N_1$, with $x_{i-1} < x_i$, $i=2,\ldots,N_1$, where Δ_1 is the distance between two successive x-observations. Also suppose that we have a regression function $y = f_1(x)$ and $Y_{1k} \sim N(\mu_1(x); \sigma_1(x))$, where $\mu_1(x) = f_1(x)$ and $\sigma_1(x) = \sigma_1/\sqrt{n_1}$. Let x_1 be an arbitrary x-observation. Then we can rewrite the x-observations as $x = x_1 + k\Delta_1$ and the mean of the distribution for Y_1 as $\mu_1(x) = f_1(x_1 + k\Delta_1)$.

Now we can generalize the simulation results obtained from this study to another corresponding situation $Y_{2k} \sim N(\mu_2(x); \sigma_2(x))$, where

$$\mu_2(x) = (\frac{\sigma_2}{\sigma_1})\sqrt{\frac{n_1}{n_2}}\ \mu_1(x).$$

If for instance $\mu_1(x)$ and $\mu_2(x)$ are analogous simple second degree polynoms, ie $\mu_i(x) = -c_i x^2$, and x_2, n_2 and Δ_2 correspond to x_1, n_1 and Δ_1, then this connection would be

$$c_1(x_1 + k\Delta_1)^2 = c_2\ (\frac{\sigma_1}{\sigma_2})\sqrt{\frac{n_2}{n_1}}\ (\frac{\Delta_2}{\Delta_1})^2\ (x_2\ \frac{\Delta_1}{\Delta_2} + k\Delta_1)^2$$

From this equality we obtain

$$\frac{c_1\Delta_1^{\,2}}{\frac{\sigma_1}{\sqrt{n_1}}} = \frac{c_2\Delta_2^{\,2}}{\frac{\sigma_2}{\sqrt{n_2}}}$$

which means that the result from the first study holds for every situation with a regression function with the same normalized curvature. We can also draw the conclusion that if the first simulation is made for the x-interval (a, b) then the generalization corresponds to the x-interval $(a\Delta_1/\Delta_2, b\Delta_1/\Delta_2)$. If we want to generalize the results in x we can argue in a corresponding way. Suppose we have the same two simulation studies described above. This means that $\mu_1(x) = -c_1x^2$ and $\sigma_1(x) = \sigma_1/\sqrt{n_1}$. We compare this first study to the other within the y-interval $-\sigma_1(x) \leq \mu_1(x) \leq 0$ of length $\sigma_1(x)$ corresponding to an x-interval $-L \leq x \leq L$ of length 2L. The other study also concerns a unimodal regression function $y_2 = f_2(x)$ with $Y_{2k} \sim N(\mu_2(x); \sigma_2(x))$. Then the following connection holds

$$Var_{II}(x_{max}) = L^2 c_2 \frac{\sqrt{n_2}}{\sigma_2} Var_I(x_{max}) \Leftrightarrow$$

$$\frac{Var_I(x_{max})}{(\Delta_1)^2} = \frac{Var_{II}(x_{max})}{(\Delta_2)^2}$$

Thus $\hat{Var}(\hat{X}_{max})$ from one study can be generalized to other situations after rescaling.

Generally we can define the normalized curvature according to the following:

Suppose we have a sample (x_i, y_{ij}), j=1,...,n, i=1,...,N, with ordered x-observations. Also suppose that we have a unimodal regression function $y = f(x)$ and $Y_k \sim N(\mu(x); \sigma/\sqrt{n})$. If $\mu(x) = -cx^2$ then we can write the definition of the normalized curvature, NC

$$NC = \frac{\mu(x_{i+1}) + \mu(x_{i-1}) - 2\mu(x_i)}{2\frac{\sigma}{\sqrt{n}}}.$$

In the simulations we use $NC = c\Delta^2\sqrt{n}/\sigma$ according to the evaluation above.

In an analogous way we can define the normalized skewness, NS

$$NS = \frac{\mu(x_{i+2}) - 3\mu(x_{i+1}) + 3\mu(x_i) - \mu(x_{i-1})}{4\frac{\sigma}{\sqrt{n}}}.$$

The normalized skewness cannot be evaluated in the same way as the normalized curvature. Thus when we examine how skewness affects bias and the variance the results from the simulation studies will generally not hold for regression functions with the same normalized skewness.

A fitting expression to be used in the simulations is

$$\frac{\mu(x_i)}{\frac{\sigma}{\sqrt{n}}} = -cx_i^2 + dx_i^3,$$

which corresponds to $NS = \dfrac{3d\Delta^3}{2\sigma/\sqrt{n}}$ and $NC = \dfrac{3d\Delta^3 i - c\Delta^2}{\sigma/\sqrt{n}}$ for $x_i = \Delta i$.

We can also conclude that when we compare isotonic regression to concave-up regression, or concave-unimodal regression to ordinary unimodal regression, and use straight lines as regression functions, then the simulation results hold for each straight line having the same normalized slope $c\Delta\sqrt{n}/\sigma$.

These kinds of expressions can of course be used together with more complicated functions used in the section concerning sigmoid regression like df(ax+b)+c.

4. SOME PROPERTIES FOR THE L.S. ESTIMATION METHOD.

4.1. Convergence of the algorithm.

The algorithm consists of an inclusion part and an exclusion part. The inclusion part works in a way that gives new bending points, one by one according to a specific criterion. After each obtained bending point the algorithm checks if the form restrictions are fulfilled before it picks another bending point. This procedure gives us an ever smaller sum of squares of errors. The inclusion part continues until the restrictions about the form are not fulfilled. Then the exclusion part starts. Thus one bending point is excluded according to a special criterion and a new solution is determined. The algorithm checks if the new solution fulfills the form restrictions. If this is not the case, another bending point is excluded. This continues until we obtain a solution which makes the restrictions fulfilled again.

We now start with a new presolution with a smaller sum of squares of errors than the previous one. Since we only have a finite number of possible sets of bending points and since each cycle gives us a smaller sum of squares of errors the procedure will converge to the LS solution in a finite number of steps.

4.2. Consistency.

Consistency in concave regression in general is a very intricate problem. We will present only one consistency property given by Hansen & Pledger (1976).

Let **I** be an interval on the real line. For each x in **I** let F_x be a cumulative distribution function with the mean $\mu(x)$. Suppose that $\mu(x)$ is continuous and concave on **I**. Also suppose that Y_j, j=1,......,N is a sequence of independent random variables which gives Y_k the distribution F_{x_k}. For each positive integer N we can define an estimator $\mu_N(x) = \mu_N(x; Y_j, j=1,......,N)$ of $\mu(x)$ which minimizes a sum of squares. In this situation with probablility one, $\mu_N(x)$ will converge to $\mu(x)$ uniformly on any subinterval of **I** and in addition $\mu_N(x)$ satisfies a condition which means that $\mu_N(x)$ will not get too large at the ends of the interval **I**.

THEOREM 4.1: Suppose that **I** = [0,1] and that 0 < a < b < 1. Then

$$P\Big\{ \limsup_{N\to\infty} \max_{x\in I} [\mu_N(x) - \mu(x)] \le 0 \,;$$

$$\liminf_{N\to\infty} \min_{a\le x\le b} [\mu_N(x) - \mu(x)] \ge 0 \Big\} = 1.$$

The proof of this theorem is given in Hanson & Pledger (1976).

4.3. Some properties of the weights of the bending points.

In some applications continuous piecewise linear regression with fixed equal sized x-intervals and equidistant bending points could be used as a reasonable regression model. In other situations this regression model can illustrate the dependence of the estimates on the length of the x-intervals. Since this is also a simplified situation we can use it to improve the understanding of the estimation method. We start by discussing some properties of the weights of the bending points using this regression model.

Suppose that (x_k, y_k), k=1,..,i, i+1,..,N, are observations with the corresponding weights w_k, in two neighbouring x-intervals, x_1-x_i and x_{i+1}-x_N. Determine the ordinary least squares estimate of the regression line, $\hat{y}_k = f(x_k) = a + bx_k$, in the first x-interval. Use the regression line to estimate a y-value for the smallest x-value in the second x-interval, $f(x_{i+1}) = a + bx_{i+1}$. The x-observation, x_{i+1}, is called the bending point of the second x-interval. This observation is denoted $(x_0^{(2)}, y_0^{(2)}) = (x_{i+1}, f(x_{i+1}))$.

According to chapter 2 the weight, $w_0^{(2)}$, of the fiction point $(x_0^{(2)}, y_0^{(2)})$ can be expressed as

$$w_0^{(2)} = \frac{\sum_{k=1}^{i} w_k \sum_{k=1}^{i} (x_k - \bar{x}_1)^2 w_k}{\sum_{k=1}^{i} (x_k - \bar{x}_1)^2 w_k + (\bar{x}_1 - x_0^{(2)})^2 \sum_{k=1}^{i} w_k}$$

where $\bar{x}_1$ denotes the weighted average of the x-observations in the first x-interval. From the last quadratic expression in the denominator we see that if $x_0^{(2)}$ is far from $\bar{x}_1$ then the denominator will be big and the weight $w_0^{(2)}$ small. It is also obvious that a regression function $f(x_k) = a + bx_k$ can be affected by the observations in a preceding x-interval and also affect the regression function in a forthcoming x-interval.

In the special case, when the first x-interval only consists of one x-observation, we obtain $w_0^{(2)} = 0$. In this situation we would skip this x-interval in the first part of the estimation procedure for the time being and start all over again in the next x-interval containing at least 2 x-observations. The $\hat{\beta}$-values that we need in the comparison in the inclusion-exclusion procedure would then be obtained in the continuous piecewise linear regression part according to the following reasoning.

Suppose we have estimated a regression line, $f_2(x) = \hat{\alpha}_2 + \hat{\beta}_2 x$, in the second x-interval, which contains at least 2 x-observations. Use this estimate to calculate the value in the bending point, $x_0^{(2)} = x_2$. The value of $\hat{\beta}_1$ is obtained by estimating a straight line through (x_2, $f_2(x_2)$) and (x_1, y_1).

If we use the assumptions of equidistant x-observations and equally sized x-intervals it is easy to evaluate some theory about $w_0^{(i)}$. A useful property describes the influence of the observations in the (i-k):th or (i+k):th x-interval on the weighted steepness in the i:th x-interval:

Suppose that (x_k, y_k), k=1,..,i, i+1,..,N, are observations with the corresponding weights w_k, and that the values of x are equidistant. In chapter 2 we described the result of the procedure, from which we obtain the estimate of the regression function, as a continuous function consisting of piecewise straight lines. In order to get rough estimates of the influence of the bounds we make a calculation for equidistant bending points. This means that in each x-interval, corresponding to an estimated straight line, we have m ordinary observations x_j, j=i,....,i+m-1, with weights $w_j = 1$ together with an extra estimated point $(x_0^{(i)}, y_0^{(i)})$ with weight $w_0^{(i)}$. This estimated observation, with corresponding weight, contains the influence of the observations in the neighbouring x-interval, i-1. This means that when we use the weighted LSE of the regression line the influence of $(x_0^{(i)}, y_0^{(i)})$ on the slope is $w_0^{(i)}/(m+w_0^{(i)})$. But, according to lemma 4.1, $w_0^{(i)}$ is limited by 0.1 and 0.3 and

$$\frac{w_0^{(i)}}{m + w_0^{(i)}} < \frac{0.3}{1 + 0.1} < 0.275.$$

Thus the influence of the observations in the (i-k):th or (i+k):th x-interval on the weighted steepness in the i:th x-interval is generally $(0.275)^k$. If $k > 4$ this influence will be very small.

Suppose we have a regression problem according to chapter 2 with observations (x_k, y_k), k=1,...,N with weights w_k. Also suppose that the x-observations are equidistant. Divide the observations into intervals, each containing m x-observations, $m \geq 2$. The first x-observation in each interval will be an ordinary observation with weight w_k and also coincide with the bending point with weight $w_0^{(i-1)}$ for the previous interval. The rest of the (m-1) x-observations in each interval will be ordinary points. In this situation we can give a general expression for the weight $w_0^{(i)}$

$$w_0^{(i)} = \frac{[w_0^{(i-1)} + \sum_{k=1}^{m} w_k][w_0^{(i-1)}(x_0-\bar{x}_{10})^2 + \sum_{k=1}^{m}(x_k-\bar{x}_{10})^2 w_k]}{\sum_{k=1}^{m}(x_k-x_0^{(i)})^2 w_k + (x_0-x_0^{(i)})^2 w_0^{(i-1)}} =$$

$$\frac{\sum_{k=1}^{m} w_k \sum_{k=1}^{m} (x_k-\bar{x})^2 w_k + [\sum_{k=1}^{m} (x_k-\bar{x})^2 w_k + \frac{\sum_{k=1}^{m} (x_k-x_0^{(i-1)})^2 w_k}{\sum_{k=1}^{m} w_k}] w_0^{(i-1)}}{\sum_{k=1}^{m} (x_k-x_0^{(i)})^2 w_k + (x_0-x_0^{(i)})^2 w_0^{(i-1)}}$$

where $\bar{x}_{10}$ is the weighted mean in the (i-1):th x-interval according to the following:

$$\bar{x}_{10} = \frac{w_0 x_0 + \sum_{k=1}^{m} w_k x_k}{w_0 + \sum_{k=1}^{m} w_k}$$

It is obvious that if the series $w_0^{(i)}$, successively obtained, will converge to the limit value, w_0, then this value is the solution of the general expression when $w_0^{(i)} = w_0^{(i-1)}$, ie

$$w_0^{(i)} \left[\sum_{k=1}^{m} (x_k-x_0^{(i)})^2 w_k + (x_0-x_0^{(i)})^2 w_0^{(i)} \right] =$$

$$\sum_{k=1}^{m} w_k \sum_{k=1}^{m} (x_k-\bar{x})^2 w_k + \left[\sum_{k=1}^{m} (x_k-\bar{x})^2 w_k + \frac{\sum_{k=1}^{m} (x_k-x_0^{(i-1)})^2 w_k}{\sum_{k=1}^{m} w_k} \right] w_0^{(i)},$$

This is a second degree equation with the solution

$$w_0 = -\frac{\sum_{k=1}^{m} (x_k-x_0) w_k + \sum_{k=1}^{m} (x_k-x_0^{(i)}) w_k}{2(x_0-x_0^{(i)})} \pm$$

$$\sqrt{\frac{\left[\sum_{k=1}^{m}(x_k-x_0)w_k+\sum_{k=1}^{m}(x_k-x_0^{(i)})w_k\right]^2-4\sum_{k=1}^{m}w_k\sum_{k=1}^{m}(x_k-\bar{x})^2w_k}{4(x_0-x_0^{(i)})^2}}$$

The general expression of the weight, $w_0^{(i)}$, can be written in a simplified way as

$$w_0^{(i)} = \frac{A + Bw_0^{(i-1)}}{C + Dw_0^{(i-1)}} = f(w_0^{(i-1)}).$$

This equation behaves very nicely. It is obvious that the size of the derivative f' will influence the distance between $w_0^{(i)}$ and w_0 and also the convergence of $\lim_{i\to\infty} w_0^{(i)}$. Generally, if f'<1 then $\lim_{i\to\infty} w_0^{(i)}$ will converge to a limit value. If on the other hand f'>1 then this is not the case. A special case is when f'=1. This means that the weights are constant ie $w_0^{(i)} = a$, i=1,...,M. We can also conclude that if the equation $(C + Dw_0^{(i-1)})w_0^{(i)} = (A + Bw_0^{(i-1)})$ has complex roots then the series $w_0^{(i)}$ will not converge. In other cases we can obtain two solutions.

In simple situations the limit value, w_0, is very easy to calculate. Let Δ be the notation of the distance between two successive x-observations. In the special case when $\Delta=1$ and $w_k=1$, k=1,...N, the solution of the general expression gives us the limit value

$$w_0 = -\frac{1}{2} \overset{+}{(-)} \sqrt{\frac{m^2+2}{12}}.$$

In this situation we can prove the convergence. But in the proof we need the relation

$$A_1*m < w_0^{(i-1)} < A_2*m \Rightarrow A_1*m < w_0^{(i)} < A_2*m.$$

Therefore in the following lemma we show that this relation holds.

LEMMA 4.1: In the regression problem described above the following holds

$$A_1*m < w_0^{(i-1)} < A_2*m \Rightarrow A_1*m < w_0^{(i)} < A_2*m.$$

PROOF: We start by determining upper and lower bounds of the limit value w_0:

$$0.03m < w_0 = -\frac{1}{2} + \sqrt{\frac{m^2+2}{12}} < 0.3m$$

In the same way we can determine the upper and lower bounds of the first $w_0^{(i)}$-value, namely $w_0^{(2)}$. After some calculations we obtain

$$w_0^{(2)} = \frac{\sum_{k=1}^{i_k} w_k \sum_{k=1}^{i_k} (x_k-\bar{x}_1)^2 w_k}{\sum_{k=1}^{i_k} (x_k-\bar{x}_1)^2 w_k + (\bar{x}_1 - x_0^{(2)})^2 \sum_{k=1}^{i_k} w_k} = \frac{m}{4}\left(1 - \frac{3}{2m+1}\right).$$

Since $m \geq 2$ we get $0.1m < w_0^{(2)} < 0.25m$. In the following $w_0^{(i)}$-series, i=3,4,....., we don´t know if the $w_0^{(i)}$-values will be less or bigger than $w_0^{(2)}$. Therefore we let the upper limit of $w_0^{(2)}$ be 0.5m and the lower limit be 0.01m. Thus w_0 will lie somewhere in the middle of the interval of $w_0^{(2)}$. Now we continue the proof by showing that

$$0.01m < w_0^{(i-1)} < 0.5m \quad \Rightarrow \quad 0.01m < w_0^{(i)} < 0.5m.$$

This is done according to the following:

In the preceding discussion we showed that

$$w_0^{(i)} = \frac{[w_0^{(i-1)} + \sum_{k=1}^{m} w_k][w_0^{(2)}(x_0-\bar{x}_{10})^2 + \sum_{k=1}^{m}(x_k-\bar{x}_{10})^2 w_k]}{w_0^{(i-1)}(x_0-x_0^{(i)})^2 + \sum_{k=1}^{m}(x_k-\bar{x}_{10})^2 w_k + (\bar{x}_{10}-x_0^{(i)})^2 \sum_{k=1}^{m} w_k}.$$

If we use the assumption of equidistant x-observations and equally sized x-intervals we obtain the following expression

$$w_0^{(i)} = w_0^{(i-1)} + \frac{\frac{m^2(m-1)(m+1)}{12} - (w_0^{(i-1)})^2 m^2 - w_0^{(i-1)} m^2}{\frac{m(m+1)(2m+1)}{6} + w_0^{(i-1)} m^2}.$$

Denote the fraction F. This gives us the following

$$F = \frac{\frac{m^2(m-1)(m+1)}{12} - (w_0^{(i-1)})^2 m^2 - w_0^{(i-1)} m^2}{\frac{m(m+1)(2m+1)}{6} + w_0^{(i-1)} m^2}$$

To find the upper and lower limits of $w_0^{(i)}$ we look at the limits of F.

$$\max\ (F \mid w_0^{(i-1)} = 0.5) < \frac{(0.084 - (0.5)^2 - 0.125)m^4}{(0.333 + 0.5)m^3} < -0.2m$$

$$\min\ (F \mid w_0^{(i-1)} = 0.01) > \frac{(0.062 - (0.01)^2 - 0.0025)m^4}{(0.6667 + 0.01)m^3} > 0.087m$$

This will give us the limits of $w_0^{(i)}$ according to

$$0.097m = (0.01+0.087)m < w_0^{(i)} = w_0^{(i-1)} + F < (0.5-0.2)m = 0.3m.$$

Thus if $0.01m < w_0^{(i-1)} < 0.5m$ then $0.01m < 0.097m < w_0^{(i)} < 0.3m < 0.5m$.

Q.E.D.

THEOREM 4.2: The weights, $w_0^{(k)}$, for the bending points, $x_0^{(k)}$, successivly obtained, will converge to the limit value

$$w_0 = -\frac{1}{2} + \sqrt{\frac{m^2 + 2}{12}}$$

PROOF: We show that the series converges by proving that the following holds

$$w_0^{(i)} - w_0 < K * (w_0^{(i-1)} - w_0) \quad \text{for } |K| < 1$$

We have already shown that the following equality holds

$$w_0^{(i)} = w_0^{(i-1)} + \frac{\frac{m^2(m-1)(m+1)}{12} - (w_0^{(i-1)})^2 m - w_0^{(i-1)} m^2}{\frac{m(m+1)(2m+1)}{6} + w_0^{(i-1)} m^2} \quad \Leftrightarrow$$

$$w_0^{(i)} - w_0 = (w_0^{(i-1)} - w_0)\left[1 - \frac{m(w_0 + w_0^{(i-1)} + 1)}{\frac{(m+1)(2m+1)}{6} + w_0^{(i-1)}m}\right].$$

Now it remains to be shown that

$$\left|1 - \frac{m(w_0 + w_0^{(i-1)} + 1)}{\frac{(m+1)(2m+1)}{6} + w_0^{(i-1)}m}\right| < 1$$

ie

$$0 < \frac{m(w_0 + w_0^{(i-1)} + 1)}{\frac{(m+1)(2m+1)}{6} + w_0^{(i-1)}m} < 2.$$

First we show that this unequality holds for $w_0^{(i-1)} = w_0^{(2)}$.
From lemma 4.1 we obtain upper and lower limits for w_0 and $w_0^{(2)}$:

$$0.03m < w_0 < 0.3m \quad \text{and} \quad 0.1m < w_0^{(2)} < 0.25m.$$

Using these limits we can calculate the upper and lower limits

$$\frac{0.11\ m^2}{0.92\ m^2} < \frac{m(w_0 + w_0^{(i-1)})}{\frac{(m+1)(2m+1)}{6} + w_0^{(i-1)}m} < \frac{0.55\ m^2}{0.43\ m^2} \Leftrightarrow$$

$$0.12 < \frac{m(w_0 + w_0^{(i-1)})}{\frac{(m+1)(2m+1)}{6} + w_0^{(i-1)}m} < 1.28 \Rightarrow$$

$$-0.28 < K < 0.89.$$

We have now obtained the desired result for i=3

$$w_0^{(3)} - w_0 < K * (w_0^{(2)} - w_0) \quad \text{for } |K| < 0.89.$$

According to lemma 4.1 $w_0^{(2)} \in I$, where $I = \{x \mid 0.01 < x < 0.5\}$ and $w_0^{(i-1)} \in I \Rightarrow w_0^{(i)} \in I$. Therefore it follows that

$$|w_0^{(i)} - w_0| < 0.89 * |w_0^{(i-1)} - w_0| \quad \text{for } i = 3,.........$$

From this unequality we obtain

$$|w_0^{(i)} - w_0| < 0.89^{(i-2)} * |w_0^{(2)} - w_0| < 0.89^{(i-2)} * 0.25m$$

which means that the series $w_0^{(i)}$ converges to the limit value. According to the discussion in the beginning of this chapter the limit value is

$$w_0 = -\frac{1}{2} + \sqrt{\frac{m^2 + 2}{12}}$$

Q.E.D.

4.4. Some general properties.

In this section we have collected the theoretical results which do not fit into any other chapter naturally. Still it is important that the results are available to the reader since they contribute to the understanding of the estimation method. The lemmas and the theorems presented here illustrate some general properties. Suppose that μ is a sigmoid function with respect to the points x_j, $j=1,...,N$, ie $y_j = \mu(x_j)$, $j=1,.....,N$ is sigmoid. Let $y_0=0$ and $x_0=0$. Also let $U=[0,1]^N$ and let **C** denote the set of every $\mathbf{y} \in U$, $\mathbf{y}=(y_1,.....,y_N)'$. Denote the line segment through (x_{j-1}, y_{j-1}) and (x_j, y_j) by

$$b_j = \frac{y_j - y_{j-1}}{x_j - x_{j-1}}, \quad j = 2,......,N.$$

We can then characterize the sigmoidicity of y in terms of linear inequality constraints. There should exist a j-value, $1 \le j \le N$, so that

$$C = \{y \in U \mid (0 \le b_2 \le ...... \le b_j \text{ and } b_j \ge b_{j+1} \ge ...... \ge b_N \ge 0\}.$$

Observe that j=1 and j=N are included as special cases. In the cases of $3 \le j \le N\text{-}1$ the interval (x_{j-1}, x_j) may naturally be called the inflection interval of the sigmoid piecewise linear function.
A very useful property is that the set of concave/convex functions is a convex cone. This is proved in the following lemma:

LEMMA 4.2: Let C_0 be a class of concave-up functions. Then this set of functions is a convex cone.

PROOF: Suppose that

$$f_k(x) = \hat{\alpha}_k^{(f)} + \hat{\beta}_k^{(f)} * x \in C_0$$

$$g_k(x) = \hat{\alpha}_k^{(g)} + \hat{\beta}_k^{(g)} * x \in C_0.$$

Thus

i) $$\hat{\beta}_k^{(f)} > \hat{\beta}_{k+1}^{(f)} \quad k=1,....,M_f$$

$$\hat{\beta}_j^{(g)} > \hat{\beta}_{j+1}^{(g)} \quad j=1,.....,M_g$$

ii) $$f(x_i) < f(x_{i+1}) \quad i=1,....,N$$

$$g(x_i) < g(x_{i+1}) \quad i=1,....,N$$

A. Then from $c \geq 0$ follows that $c*f_k(x) \in C_0$ since

i) $c*\hat{\beta}_k^{(f)} > c*\hat{\beta}_{k+1}^{(f)}$ $k=1,....,M_f$

ii) $c*f(x_i) < c*f(x_{i+1})$ $i=1,....,N.$

B. For $0 \leq a \leq 1$ we obtain that $a*f_k(x) + (1\text{-}a)*g_j(x) \in C_0$ since

i) $a*\hat{\beta}_k^{(f)} + (1\text{-}a)*\hat{\beta}_j^{(g)} > a*\hat{\beta}_{k+1}^{(f)} + (1\text{-}a)*\hat{\beta}_j^{(g)} > a*\hat{\beta}_{k+1}^{(f)} + (1\text{-}a)*\hat{\beta}_{j+1}^{(g)}$

or

$a*\hat{\beta}_k^{(f)} + (1\text{-}a)*\hat{\beta}_j^{(g)} > a*\hat{\beta}_k^{(f)} + (1\text{-}a)*\hat{\beta}_{j+1}^{(g)} > a*\hat{\beta}_{k+1}^{(f)} + (1\text{-}a)*\hat{\beta}_{j+1}^{(g)}$

ii) $a*f(x_i) + (1\text{-}a)*g(x_i) < a*f(x_{i+1}) + (1\text{-}a)*g(x_{i+1}).$

If these two conditions are fulfilled C_0 is a convex cone.
The situations when C_0 is the class of convex-up, concave-down or convex-down functions are proved in an analogous way.

Q.E.D.

It is easy to expand this proof to hold for the situations when
C_0 is the class of unimodal functions with maximum in a common fixed $x = x_{max}$,
C_0 is the class of sigmoid functions with inflection point in a common fixed $x = x_{inf}$.

The next theorem deals with the inflection interval in sigmoid regression. Suppose that we have used the estimation method described in section 2 and obtained an estimate of the sigmoid regression function. If the obtained estimate is the least squares solution the length of the inflection interval is determined by the distance between two successive x-observations.

THEOREM 4.3: The least squares estimate of a sigmoid regression function has an inflection interval between two neighbouring x-observations with the rare exception when all inner points of the inflection interval have 0 error.

PROOF: Suppose that we have obtained a solution where the inflection interval, ie the k:th x-interval contains several x-observations. These observations are denoted by (x_j, y_j), $j = i_{(k-1)}+1, \ldots\ldots, i_k$, and the corresponding estimated inflection interval by $f_k^*(x_j)$, $j = i_{(k-1)}+1, \ldots\ldots, i_k$. Observe that the end observations of the inflection interval, $y_{i_{(k-1)}+1}$ and y_{i_k}, include fiction points with influence from the surrounding x-intervals. If we examine the observation (x_{i_k}, y_{i_k}) we will find that it will lie above the estimated regression line $f_k^*(x)$. Denote the residuals corresponding to the estimated inflection interval $\varepsilon_j = y_j - f_k^*(x_j)$, for $j = i_{(k-1)}+1, \ldots, i_k$. For each j^*, $i_{(k-1)}+1 \le j^* \le i_k$, we can define a function

$$
v_{j^*}(x) = \begin{cases}
0 & \text{for } x \le x_{i_{(k-1)}+1} \\[6pt]
\dfrac{x - x_{i_{(k-1)}+1}}{x_{j^*} - x_{i_{(k-1)}+1}} & \text{for } x_{i_{(k-1)}+1} < x \le x_{j^*} \\[6pt]
\dfrac{x_{i_k} - x}{x_{i_k} - x_{j^*}} & \text{for } x_{j^*} < x \le x_{i_k} \\[6pt]
0 & \text{for } x_{i_k} < x
\end{cases}
$$

For some j^* we must have

$$\sum_{\nu=i_{(k-1)}+1}^{i_k} v_{j^*}(x_\nu)\varepsilon_\nu > 0 \quad \text{or} \quad \varepsilon_\nu \equiv 0$$

for $\nu = i_{(k-1)}+1,\ldots\ldots,i_k$. This is due to the fact that $f_k^*(x_j)$ can be determined as an ordinary linear regression of the observations in the estimated inflection interval, including the fiction points, and that the set $v_{j^*}(x)$, $j^*=i_{(k-1)}+1,\ldots\ldots,i_k$, is a complete function system on $x_{i_{(k-1)}+1},\ldots\ldots,x_{i_k}$.If

$$\sum_{\nu=i_{(k-1)}+1}^{i_k} v_{j^*}(x_\nu)\varepsilon_\nu > 0$$

then for small $\delta > 0$ we can find a new estimate of the regression function using $f_k^*(x_j) + \delta v_{j^*}(x_j)$. This estimate will also be a sigmoid curve for a small enough δ with the sum of squares of errors:

$$\sum_{\nu=i_{(k-1)}+1}^{i_k} [y_\nu - (f_k^*(x_\nu)+\delta v_{j^*}(x_\nu))]^2 = \sum_{\nu=i_{(k-1)}+1}^{i_k} [y_\nu - f_k^*(x_\nu)]^2 -$$

$$2\delta \sum_{\nu=i_{(k-1)}+1}^{i_k} [f_k^*(x_\nu)v_{j^*}(x_\nu)]^2 + \delta^2 \sum_{\nu=i_{(k-1)}+1}^{i_k} [v_{j^*}(x_\nu)]^2$$

If δ is chosen small enough we get the unequality

$$\delta^2 \sum_{\nu=i_{(k-1)}+1}^{i_k} [v_{j^*}(x_\nu)]^2 \ - \ 2\delta \sum_{\nu=i_{(k-1)}+1}^{i_k} [f_k^{\,*}(x_\nu)v_{j^*}(x_\nu)]^2 \ > \ 0$$

ie

$$\sum_{\nu=i_{(k-1)}+1}^{i_k} [y_\nu - f_k^{\,*}(x_\nu)]^2 \ > \ \sum_{\nu=i_{(k-1)}+1}^{i_k} [y_\nu - (f_k^{\,*}(x_\nu)+\delta v_{j^*}(x_\nu))]^2.$$

Thus we have found another sigmoid estimate with a smaller sum of squares of errors and with a shorter inflection interval than before. This argument can be used successively in the inflection interval until we only have inflection interval between two neighbouring x-observations left.

Q.E.D

Summarizing the estimation method we can conclude that the estimate of concave/convex regression is a piecewise linear function. However, the formulas for continuous piecewise regression function estimates are not directly applicable to this case. But the sum of the mean variance and the mean squared bias for piecewise linear regression is probably a rough estimate of the mean variance of a concave/convex estimate with the same mean interval length. One common property for continuous piecewise linear regression is that the least squares regression function estimates are linear functions of the observations. This is proved in the following lemma.

LEMMA 4.3: If the bending points are fixed then the LS regression functions estimates are linear functions of the observations.

PROOF: For any of the x_i-intervals where the function is linear, the estimates are determined by using the estimation method in step 1 simultaneously from the first x-interval to the i:th and from the last x-interval to the i:th. Then information from the observations in the first interval to the (i-1):th is contained in a fiction point in the i:th bending point and information from the (i+1):th to the last x-interval is contained in a fiction point in the (i+1):th bending point. The regression function estimate in the i:th x-interval is the ordinary linear regression function estimate with the two fiction points included and for each observation in the interval the estimate is a linear function of all observations. But the fiction points are also a linear function of the observations in the other intervals respectively to the left of the (i-1):th and to the right of the (i+1):th intervals including other fiction points and so on. Therefore we can conclude that each regression function estimate is a linear function of the observations.

Q.E.D.

Suppose that we have a sample y_{ij}, $j=1,...,n(x_i)$, $i=1,...,N$, where $E(Y_{ij}) = \mu(x_i)$ is a concave function and $Y_{ij} \sim N(\mu(x_i); \sigma(x_i))$. Using some local maxima we can form a concave function, $g_N(x_i)$, which is an upper boundary of the set of observations. It is now obvious that $\min(\hat{Y}(x_i), g_N(x_i))$ is also concave and this function would have a smaller LS distance to the observations than $\hat{Y}(x_i)$ if $\hat{Y}(x_i) > g_N(x_i)$ in some point. This can not happen since $\hat{Y}(x_i)$ is the LS concave estimate. This means that $g_N(x_i) > \hat{Y}(x_i)$ for $i=1,...,N$, is an upper limit of the LS concave regression function, $\hat{Y}(x_i)$, for all observation points x_i. The inequality can be used in order to get bounds of statistic properties of the LS concave estimates.

5. BIAS.

5.1. Bias obtained from the estimation method

5.1.1. Bias of $\hat{Y}(x_k)$ for a related problem.

Since the least squares estimation method is rather complicated, we first examine what bias the estimation method will give in an interval without any fiction point. In chapter 2 we gave the following definition of a fiction point.

> DEF. A predicted point, $(x_0^{(k)}, y_0^{(k)})$, where $x_0^{(k)}$ is the bending point of the k:th x-interval and $y_0^{(k)}$ is the corresponding y-value, predicted according to the description, is called the fiction point from the k:th x-interval.

Suppose we have the observations $(x_1, y_1), (x_2, y_2), \ldots\ldots, (x_m, y_m)$ with $d_1 \le x_1 < x_2 \ldots\ldots < x_m \le d_2$, where the x-observations are equidistant with the distance Δ between two successive x-observations and with the non-negative weights $w_1, w_2, \ldots, w_m$. Then the number of x-observations is

$$m = \frac{(d_2 - d_1)}{\Delta} + 1$$

Also suppose that we have a concave-unimodal regression function, for instance $f(x)=a-cx^2$, in the interval $d_1 \le x \le d_2$. The estimation procedure will estimate the regression function as a weighted least squares regression line. The expectation is

$$E(Y_k) = a - \frac{c\sum_{i=1}^{m} x_i^2 w_i}{\sum_{i=1}^{m} w_i} + b\,(x_k w_k - \bar{x})$$

where

$$b = \frac{-c\sum_{i=1}^{m}[x_i - \bar{x}]x_i^2 w_i}{\sum_{i=1}^{m}[x_i - \bar{x}]^2 w_i} \quad \text{and} \quad \bar{x} = \frac{\sum_{i=1}^{m} x_i w_i}{\sum_{i=1}^{m} w_i}$$

Bias, which is defined as $Y_{bias}(x_i) = E(Y_i) - f(x_i)$, will of course vary with the x-observations. From the definition we get

$$Y_{bias}(x_k) = \frac{-c\sum_{i=1}^{m} x_i^2 w_i}{\sum_{i=1}^{m} w_i} + \frac{-c\sum_{i=1}^{m}[x_i - \bar{x}]x_i^2 w_i}{\sum_{i=1}^{m}[x_i - \bar{x}]^2 w_i}(x_k w_k - \bar{x}) + c x_k^2$$

Using this expression it is easy to calculate bias for different values of c, m, d_1 and d_2 and for different weight functions. Let c=1 and let the x-interval be symmetric around x=0, ie $-d_1=d_2=d$. The least squares estimate of the regression line for these x-observations will then be

$$E(Y_i) = \frac{-\sum_{i=1}^{m} x_i^2}{m} = -\frac{d^2}{3}\frac{(m+1)}{(m-1)}.$$

Bias will be obtained as $Y_{bias}(x_i) = x_i^2 - d^2(m+1)/3(m-1)$, which has a minimum for $x_i=0$.

$Y_{bias}(x_i)$ has the roots $-d\sqrt{(m+1)/3(m-1)}$ and $d\sqrt{(m+1)/3(m-1)}$ which means that $Y_{bias}(x_i)$ is negative within the interval

$$-d\sqrt{\frac{1}{3}\frac{(m+1)}{(m-1)}} < x < d\sqrt{\frac{1}{3}\frac{(m+1)}{(m-1)}}.$$

It also obvious that if x_i are equidistant there will be more x-observations having negative bias than x-observations with positive bias since the distance between the $x = 0$ and $x = d\sqrt{(m+1)/3(m-1)}$ is longer than between $x = d\sqrt{(m+1)/3(m-1)}$ and $x = d$.

It is also easy to see that bias in negative direction will tend to be smaller then bias in positive direction. This is done by comparing the distance between $y= -d^2(m+1)/3(m-1)$ and $y = 0$ with the distance between $y= -d^2(m+1)/3(m-1)$ and $y = -d^2$:

$$Y_{bias}(x=d) - Y_{bias}(x=0) = d^2 - \frac{2d^2}{3}\frac{(m+1)}{(m-1)} = \frac{d^2}{3}\frac{(m-5)}{(m-1)}.$$

From this equation we can see that if we have x-observations on a concave-unimodal regression function and the number of equidistant x-observations is m, then for

m=2 there will be no bias since the estimation procedure will estimate a regression line through the observations themselves,

m=3 or **m=4** bias with positive sign tends to be smaller than bias with negative sign,

m≥5 the size of the largest bias with positive sign is at least as big as the size of the largest bias with negative sign.

We can also roughly conclude the result by saying:

If $m \geq 5$ we will have many x-observations in the inner interval with a rather small bias with a negative sign and few x-

observations in the end intervals with bigger bias with a positive sign. This tendency will be more striking as m grows.
The positive bias at both ends of the x-interval and the negative bias in the middle indicate that the average bias is close to zero. However for the whole regression function we will have the average bias = 0. This is due to the fact that for each value of the constants a_0 and a_1 we can form the function $h(x) = a_0 + a_1 x$ such that

$$\sum_{i=1}^{N} h(x_i)[Y(x_i)-\hat{Y}(x_i)] = 0.$$

The above calculations show what happens in an x-interval without a fiction point. But since the estimation procedure is symmetric we can choose where to start

1) from the first x-interval to the last and then back again to the first x-interval,
2) from the last x-interval to the first and then back to the last x-interval,
3) from both end intervals at the same time to the middle and then back to the end intervals.

This means that the calculations also show the tendencies of the estimates obtained from the proposed estimation procedure.

5.2. Bias of $\hat{Y}(x_k)$ for isotonic concave-up regression functions.

From the calculations in chapter 5.1 we can see that the estimate of $\mu(x_k)$ will probably have a small bias for concave regression. To examine the nature of bias we start by performing two simulation studies under the restriction that the regression function is concave and increasing. The first study deals with regression functions having constant curvature. The other study deals with regression functions with linear functions in order to compare the obtained bias coursed by the estimation method in different situations.

5.2.1. Bias of $\hat{Y}(x_k)$ for isotonic-concave regression functions with constant curvature.

A simulation study was performed to examine what bias we would get for arbitrary points on the regression curve using the least squares estimation method. We also wanted to know how curvature influenced bias.

Suppose we have a sample $y_k = \mu(x_k) + \varepsilon_k$, k=1,....,N, where $\mu(x_k)$ is an increasing and concave regression function with equidistant x-observations in the interval $-d \leq x \leq 0$ and $\varepsilon_k \sim N(0, \sigma(x_k)^2)$, where $\sigma(x_k)^2$ are bounded. The regression functions examined were of the type $\mu(x)=-cx^2$ for c in the interval $0.05 \leq c \leq 8.0$.

In the simulations with the regression function $\mu(x) =-cx^2$ the normalized curvature (NC) can be expressed in terms of the curvature (c), the distance between two successive x-observations (Δ), the variance (σ^2) and the number of y-observations (n) for each x_k as NC = $\frac{c\Delta^2}{\sigma/\sqrt{n}}$. The simulation study was performed for

1) 15 equidistant x-observations in the interval $-7 \leq x \leq 0$, ie the distance between two successive x-observations was Δ=0.5.
2) $n(x_k)=1$, k=1,...,15.
3) $\varepsilon_k \sim N(0, 1)$.

The result from 25 000 replications shows $\hat{Y}_{bias}(x_k)$, k=1,...,15, in table 5.1. Since the tendency of bias was the same for all examined values of the normalized curvature we only illustrate the result for NC = 0.25, 0.50 and 0.75. The standard deviations of the means in the simulations lie between 0.004 - 0.006, estimated by the standard deviations in the simulations.

TABLE 5.1. $NC = \frac{c\Delta^2}{\sigma/\sqrt{n}}$

x_i	0.25	0.50	0.75
-7.0	-0.169	-0.108	-0.074
-6.5	0.107	0.008	0.066
-6.0	0.063	0.028	0.014
-5.5	0.013	-0.003	-0.004
-5.0	-0.003	0.001	0.004
-4.5	-0.007	-0.002	-0.002
-4.0	-0.011	-0.011	-0.011
-3.5	-0.011	-0.013	-0.014
-3.0	-0.002	0.000	0.001
-2.5	-0.002	-0.002	-0.002
-2.0	-0.006	-0.006	-0.008
-1.5	-0.008	-0.006	-0.005
-1.0	-0.022	-0.010	-0.008
-0.5	-0.045	-0.054	-0.051
0.0	0.058	0.053	0.047

Table 5.1. Random numbers were generated so that $Y \sim N(-cx_i^2;1)$ for 15 equidistant x-observations in the x-interval $-7 \leq x \leq 0$. Bias of $\hat{Y}_i$ was determined for different values of the normalized curvature, NC.

From the table we can see that bias varies slowly. It is also obvious that the estimation procedure gives larger bias at both ends of the interval $-7 \leq x \leq 0$ than in the middle. If we cut off a few observations at each end we get an estimate with negligible bias for the rest of the observations. From section 3.2 follows that this result will hold also for every regression function of the form $y = -NCx^2$ in the corresponding x-interval $-7/2\Delta_2 \leq x \leq 0$. From the table we can also see that the curvature doesn´t seem to influence bias except at also see that the curvature doesn´t seem to influence bias except at the end points where bias gets smaller when the curvature increases.

5.2.2. Bias of $\hat{Y}(x_k)$ for isotonic-concave regression functions with linear functions.

Another simulation study was performed to examine how bias was affected if the curvature was 0. To be able to compare the simulation results in this study to the results in the preceding study with a constant curvature an analogous situation was examined. Suppose we have a sample $y_k = \mu(x_k) + \varepsilon_k$, k=1,....,N, where $\mu(x_k)$ is an isotonic and concave regression function with equidistant x-observations in the interval $-d \le x \le 0$ and $\varepsilon_k \sim N(0, \sigma(x_k)^2)$, where $\sigma(x_k)^2$ are bounded. The regression functions examined were of the type $\mu(x) = cx$ for some different c-values from 0.25 to 3.0. In the simulations with the regression function $\mu(x) = cx$, the normalized slope, $N\beta$, can be expressed in terms of the distance between two successive x-observations (Δ), the variance (σ^2) and the number of y-observations (n) for each x_k as

$$N\beta = \frac{c\Delta}{\sigma/\sqrt{n}}$$

The simulation study was performed for

1) 21 equidistant x-observations in the interval $-20 \le x \le 0$, ie the distance between two successive x-observations was $\Delta=1.0$.
2) $n(x_k)=1$, k=1,...,21.
3) $\varepsilon_k \sim N(0, 1)$.

The result from 25 000 replications for Nβ = 0.25, 0.50 and 0.75 is shown in table 5.2.

TABLE 5.2. $N\beta = \frac{c\Delta}{\sigma/\sqrt{n}}$

x_i	0.25	0.50	0.75
-20.0	-0.504	-0.505	-0.506
-19.0	-0.159	-0.161	-0.162
-18.0	-0.046	-0.048	-0.049
-17.0	0.018	0.015	0.014
-16.0	0.059	0.057	0.056
-15.0	0.089	0.088	0.087
-14.0	0.111	0.110	0.109
-13.0	0.126	0.126	0.125
-12.0	0.135	0.136	0.136
-11.0	0.140	0.141	0.141
-10.0	0.139	0.142	0.143
-9.0	0.134	0.138	0.140
-8.0	0.125	0.131	0.132
-7.0	0.111	0.118	0.121
-6.0	0.090	0.101	0.104
-5.0	0.062	0.076	0.081
-4.0	0.024	0.043	0.049
-3.0	-0.026	-0.005	0.004
-2.0	-0.095	-0.077	-0.065
-1.0	-0.186	-0.194	-0.196
0.0	-0.315	-0.397	-0.439

Table 5.2. Random numbers were generated so that $Y \sim N(\mu(x);1)$ for 21 equidistant x-observations in the x-interval $-20 \le x \le 0$, where $\mu(x)=cx$. Bias of $\hat{Y}_i$ was determined for different values of the normalized slope, Nβ.

From the table we see that bias of the central x-observations are almost the same for the three constants. This is logical if the slope is not too small, since then it is only the condition of concavity that will influence the results in this area. In this simulation study we also examine how big the fractions between bias and the corresponding expected y-values are, compared to the situation with constant curvature. In all cases with linear functions bias is bigger than for situations with constant curvature. For linear functions the simulation results also shows a more clear tendency of either increasing bias or decreasing bias, depending on where the x-observation is placed in the examined x-interval. The standard deviations of the means in the simulations lie between 0.0016 and 0.0042.

5.3. Bias of $\hat{Y}_{max}$ for concave-unimodal regression functions.

In many situations the maximum point is the most interesting one. Therefore this section will deal with problems concerning bias of the estimated maximum point. We will also try to illustrate the nature of the bias. This is done both by theoretical calculations and by using some simulations results for $\hat{Y}_{max}$ and the largest ordered $\hat{Y}$-observations.

5.3.1. The influence on bias of $\hat{Y}_{max}$ by the curvature of the regression function.

When we deal with unimodal regression functions, the bias of $\hat{Y}_{max}$ can be affected by some properties of the regression functions. One of the properties we will examine is the influence on bias by the curvature. In this situation we can distinguish two separate cases: regression functions with constant curvature and regression functions with linear functions. Two simulation studies have been performed to see if bias is affected in different ways by these cases.

5.3.1.1. Bias of $\hat{Y}_{max}$ for concave-unimodal regression functions with constant curvature.

Suppose we have a sample $y_{ij} = \mu(x_i) + \varepsilon_{ij}$, j=1,....,$n(x_i)$, for each of N fixed equidistant x-values, where $\mu(x_i)$ is a symmetric unimodal regression function with maximum in x = 0. Let $\varepsilon_i \sim N(0, \sigma(x_i)^2)$, where $\sigma(x_i)^2$ are bounded.

To examine how the curvature of the regression function influences bias of $\hat{Y}_{max}$ a simulation study was performed. Bias is defined as $\text{Bias}(\hat{Y}_{max}) = \hat{Y}_{max} - \mu(x_{max})$. The regression functions examined were of the type $\mu(x) = -cx^2$ for the variation interval $0.05 \leq c \leq 3.25$. The corresponding x-interval was $-10 \leq x \leq 10$. The position of the nearest x-value was uniformly distributed in the interval $(-0.5\Delta, 0.5\Delta)$ centered at x_{max}, where Δ is the distance between two successive x-observations. The result from 25 000 replications is shown in table 5.3.

TABLE 5.3.

$NC = \frac{c\Delta^2}{\sigma/\sqrt{n}}$	$\text{Bias}(\hat{Y}_{max})$	$NC = \frac{c\Delta^2}{\sigma/\sqrt{n}}$	$\text{Bias}(\hat{Y}_{max})$
0.05	0.227	1.25	0.155
0.10	0.244	1.50	0.123
0.20	0.258	1.75	0.089
0.30	0.261	2.00	0.059
0.40	0.258	2.25	0.021
0.50	0.239	2.50	-0.013
0.75	0.215	2.75	-0.047
1.00	0.186	3.00	-0.080

Table 5.3. Random numbers were generated so that $Y \sim N(-cx_i^2;1)$ for 21 equidistant x-observations in the x-interval $-10 \leq x \leq 10$. The position of the nearest x-value was uniformly distributed in the interval $(-0.5\Delta, 0.5\Delta)$ centered at x_{max}. Bias of $\hat{Y}_{max}$ was determined for $0.05 \leq NC \leq 3.00$.

From this table we see that for values of the normalized curvature in the interval

$$1.75 \le NC = \frac{c\Delta^2}{\sigma/\sqrt{n}} \le 2.75$$

$\hat{Y}_{max}$ will have a rather small bias. It also shows that bias increases up to a maximum near NC = 0.30 and then decreases again. For $0.5 \le NC \le 3.00$ bias increases if c and $\sigma/\sqrt{n}$ remain constant and the distance, Δ, between two successive x-observations increases. The standard deviations of the means in the simulations lie between 0.003 and 0.006.

The simulation results are also shown in figure 5.1.

FIGURE 5.1.

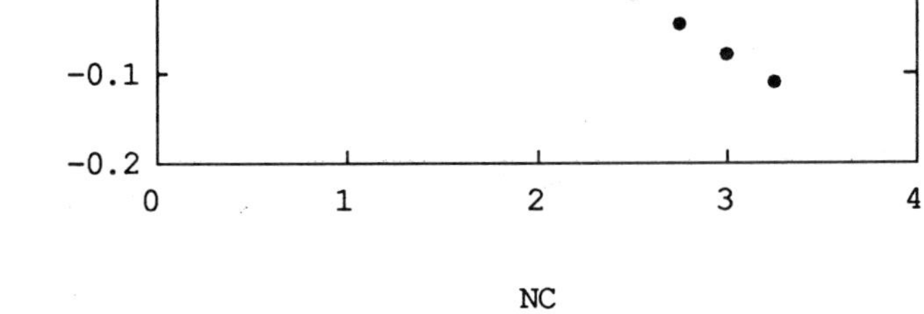

Fig 5.1. Random numbers were generated so that $Y \sim N(-cx_i^2;1)$ for 21 equidistant x-observations in the x-interval $-10 \le x \le 10$. The position of the nearest x-value was uniformly distributed in the interval $(-0.5\Delta, 0.5\Delta)$ centered at x_{max}. Bias of $\hat{Y}_{max}$ was determined for different values of the normalized curvature, NC.

To examine what the limit value of the bias is when $c \to 0$, another simulation study was performed for a wider x-interval, $-25 \le x \le 25$. The choice of this interval was due to the fact that only y-observations that lie closer than -25/c to y_{max} will have an essential influence on the estimation of y_{max}. The variation interval for c was $0.005 \le c \le 0.055$. The result from 25 000 replications is shown in table 5.4.

TABLE 5.4.

$NC = \frac{c\Delta^2}{\sigma/\sqrt{n}}$	$Bias(\hat{Y}_{max})$	$NC = \frac{c\Delta^2}{\sigma/\sqrt{n}}$	$Bias(\hat{Y}_{max})$
0.005	0.142	0.030	0.204
0.010	0.150	0.035	0.222
0.015	0.165	0.045	0.251
0.020	0.178	0.050	0.258
0.025	0.189	0.055	0.263

Table 5.4. Random numbers were generated so that $Y \sim N(-cx_i^2;1)$ for 51 equidistant x-observations in the x-interval $-25 \le x \le 25$. The position of the nearest x-value was uniformly distributed in the interval (-0.5Δ, 0.5Δ) centered at x_{max}. Bias of $\hat{Y}_{max}$ was determined for $0.005 \le NC \le 0.055$.

The standard deviations of the means in the simulations lie between 0.021 and 0.034.

From this study and the earlier study we see that bias will be affected by observations more distant from y_{max} than -25/c. Even if the affect is not essential the estimates will differ in the second decimal if the x-interval is changed from $-10 \le x \le 10$ to $-25 \le x \le 25$. For instance for NC = 0.05 $Bias(\hat{Y}_{max})$ is estimated by 0.227 for the examined x-interval $-10 \leq x \leq 10$ and by 0.258 for the examined x-interval $-25 \leq x \leq 25$. The limit value of Bias($\hat{Y}_{max}$) is the value that we would obtain if NC = 0, ie if we have a straight line parallell to the x-axis. In chapter 5.2.2 we have examined bias of $\hat{Y}(x_k)$ for linear functions. The examined linear functions are not parallell to the x-axis but the normalized slope is not too big. Another difference is that in chapter 5.2.2 bias is examined for fixed x-observations while in table 5.4 bias is obtained for the estimated maximum point, for which the x-value is not fixed. But the two estimation problems are still rather similar and therefore fairly comparable. Thus Bias($\hat{Y}_{max}$) = 0.142 obtained for NC = 0.005 in table 5.4 can be compared to bias in table 5.2 obtained from the x-observations in the middle of the examined x-interval.

In chapter 7.1.1 a limit distribution, G, containing $\mu(x)$ and $\hat{\mu}(x)$ has been examined for fixed x-values. It is supposed that the observation points are chosen in such a way that their distribution converges to F(x), which is differentiable. In this situation Mammen (1988) shows that

$$C* = N^{2/5} \frac{F'(x_0)^{2/5}}{\sigma(x_0)^{4/5}\,\mu''(x_0)^{1/5}} \left(\hat{\mu}(x_0) - \mu(x_0) \right) \rightarrow G$$

for concave regression, where

$\mu''(x_0)$ = the curvature in x_0,
$\sigma(x_0)^2$ = Var(Y|X=x_0) and
N $F'(x_0)$ = the density of observations in x_0.

This distribution is not exactly applicable on Bias($\hat{Y}_{max}$) since $\hat{X}_{max}$ is not fixed. However, we can give a rather close approximation of bias in the estimated maximum point when the normalized curvature, NC, is small. Since this is a rough approximation we use the relation C* obtained in chapter 7.1.1 for an x-observation, x= -2.600, in the middle of the examined x-interval. In this situation we obtain C* = 0.3142/2NC. The approximations for different values of NC are shown in table 5.5 together with the estimates shown in table 5.4 obtained from 25 000 replications.

TABLE 5.5.

NC	Bias($\hat{Y}_{max}$)	approximation of bias
0.005	0.142	0.125
0.010	0.150	0.144
0.015	0.165	0.156
0.020	0.178	0.165
0.025	0.189	0.173
0.030	0.204	0.179
0.035	0.222	0.185
0.040	0.235	0.190
0.045	0.251	0.194
0.050	0.258	0.198
0.055	0.263	0.202

Table 5.5. An approximation of Bias($\hat{Y}_{max}$) using the theory from Mammen (1988) is shown for different values of NC together with the estimates of Bias($\hat{Y}_{max}$) shown in table 5.4.

In this table we can see that the approximation underestimates bias in the maximum point. This is logical since the approximation is made for a fixed x-value while the estimated x_{max} varies and the maximum point is obtained as the largest predicted y-value. The results from the last simulation study are also shown in figure 5.2. This figure also indicates that Bias($\hat{Y}_{max}$) will converge to a limit value > 0 when the normalized curvature decreases to zero.

FIGURE 5.2.

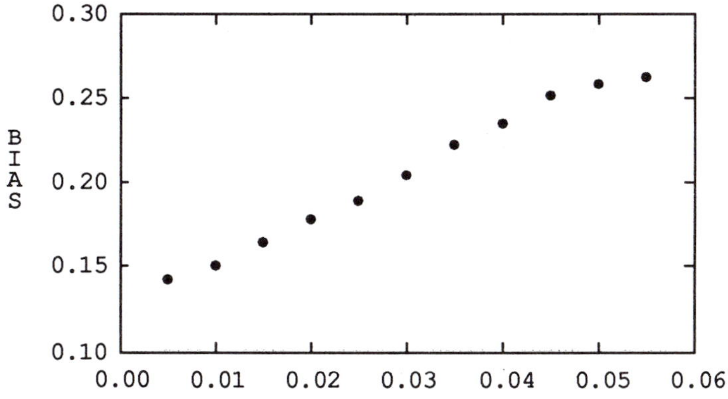

Fig 5.2. Random numbers were generated with distribution $Y \sim N(-cx_i^2;1)$ for 51 equidistant x-observations in the x-interval $-25 \leq x \leq 25$. The position of the nearest x-value was uniformly distributed in the interval $(-0.5\Delta, 0.5\Delta)$ centered at x_{max}. Bias($\hat{Y}_{max}$) was determined for different values of c.

5.3.1.2. Bias of $\hat{Y}_{max}$ for symmetric regression functions with piecewise linear functions.

To be able to compare this simulation study using piecewise linear regression functions with the preceding we used a situation as similar as possible. In this study we also had N fixed equidistant x-values, where $\mu(x_i)$ was a symmetric unimodal regression function with maximum in x = 0. The regression functions examined were of the type $y_{ij} = \mu(x_i) + \varepsilon_{ij}$, $j=1,....,n(x_i)$, where $\varepsilon_i \sim N(0, \sigma(x_i)^2)$ for bounded $\sigma(x_i)^2$. In the simulations we use

$$\mu(x) = \begin{cases} cx & \text{for } x \leq 0 \\ -cx & \text{for } x > 0 \end{cases}$$

for the variation interval $0.50 \leq c \leq 2.00$. The corresponding x-interval was $-25 \leq x \leq 25$. The position of the nearest x-value was uniformly distributed in the interval $(-0.5\Delta, 0.5\Delta)$ centered at x_{max}. The result from 25 000 replications is shown in table 5.6.

TABLE 5.6.	$N\beta = \dfrac{c\Delta}{\sigma/\sqrt{n}}$	**Bias($\hat{Y}_{max}$)**
	0.50	-0.467
	0.75	-0.574
	1.00	-0.667
	1.25	-0.749
	1.50	-0.825
	1.75	-0.897
	2.00	-0.967

Table 5.6. Random numbers were generated so that $Y \sim N(\mu(x);1)$ for 51 equidistant x-observations in the x-interval $-25 \leq x \leq 25$, where $\mu(x)=cx$ for $x<0$ and $\mu(x)=-cx$ for $x\geq 0$. The bias of $\hat{Y}_{max}$ was estimated for different values of the normalized slope, $N\beta$.

From this table we can see that for the examined values of the normalized slope, $N\beta$, all estimates of y_{max} were negatively biased. The bias also became more and more negative as the normalized slope was growing bigger. The standard deviations of the means in the simulations lie between 0.0024 - 0.0035.

5.3.2. The influence on bias of $\hat{Y}_{max}$ by the skewness of the regression function.

In the same way as the influence by the curvature on bias was examined for two different situations, this problem is discussed for two analogous functions.

5.3.2.1. The influence by the skewness on bias of $\hat{Y}_{max}$ when the regression function is a third-degree polynomial.

In this simulation study regression functions of the form $\mu(x) = -cx^2 + dx^3$ were used in the x-interval [-10, 10] for $0.8 \le c \le 3.0$ and for $0.00 \le d \le 0.05$. The x-observations were equidistant and the position of the nearest x-value was uniformly distributed in the interval $(-0.5\Delta, 0.5\Delta)$ centered at x_{max}.

TABLE 5.7.

	d		
c	**0.000**	**0.025**	**0.050**
0.8	0.200	0.198	0.203
0.9	0.188	0.187	0.192
1.0	0.186	0.176	0.173
1.5	0.123	0.113	0.111
2.0	0.059	0.045	0.044
2.5	-0.013	-0.024	-0.024
3.0	-0.080	-0.095	-0.092

Table 5.7. Random numbers were generated so that $Y \sim N(\mu(x);1)$ for 21 equidistant x-observations in the x-interval $-10 \le x \le 10$, where $\mu(x) = -cx^2 + dx^3$. The bias of $\hat{Y}_{max}$ was estimated for $0.8 \le c \le 3.0$ and $0.00 \le d \le 0.05$.

The choice of the variation intervals for c and d was dependent on the fact that $\mu(x)$ is a third-degree function, which has a minimum in $x = 2c/3d$. For bigger x-values we have no longer a unimodal function. The chosen values of c and d give us this minimum outside the examined x-interval [-10, 10]. The size of the bias of $\hat{Y}_{max}$ for $\sigma=1$, $\Delta=1$ and $n=1$ using 25 000 replications is shown in table 5.7. The standard errors of the means in the simulations lie between 0.014 - 0.020. From the simulation results we can see that the skewness was too small compared to the curvature to have any essential influence on bias.

5.3.2.2. The influence by the skewness on bias of $\hat{Y}_{max}$ for piecewise linear regression functions.

In order to examine bias of $\hat{Y}_{max}$ for piecewise linear regression functions an analogous simulation study was performed for the regression functions

$$\mu(x) = \begin{cases} cx & \text{for } x \leq 0 \\ -dx & \text{for } x > 0 \end{cases}$$

where $Y_i = \mu(x_i) + \varepsilon_i$ and $\varepsilon_i \sim N(0;1)$ in the interval $-25 \leq x \leq 25$. The values of the constants were $d = 0.5$, 1.0 and 1.5 while c varied from 0.5 to 3.0. In the same way as in the preceding study the x-observations were chosen equidistantly. The position of the nearest x-value was uniformly distributed in the interval $(-0.5\Delta, 0.5\Delta)$ centered at x_{max}. The simulation results from 25 000 replications are shown in table 5.8. The standard errors of the means in the simulations lie between 0.002 - 0.004. From the simulation results we can see that bias is negative for all values of c and d. Bias becomes more and more negative when c and/or d increase.

TABLE 5.8.

	d		
c	**0.50**	**1.00**	**1.50**
0.5	-0.463	-0.528	-0.554
1.0	-0.532	-0.667	-0.723
1.5	-0.559	-0.726	-0.818
2.0	-0.573	-0.766	-0.880
3.0	-0.587	-0.815	-0.961

Table 5.8. Random numbers were generated so that $Y \sim N(\mu(x);1)$ for 51 equidistant x-observations in the x-interval $-25 \leq x \leq 25$, where $\mu(x)=cx$ for $x \leq 0$ and $\mu(x)=-dx$ for $x>0$. The bias of $\hat{Y}_{max}$ was estimated for $0.5 \leq c \leq 3.0$ and $0.5 \leq d \leq 1.5$.

5.3.3. Illustration of the nature of bias of $\hat{Y}_{max}$ by theoretical calculations and by simulations.

As we concluded in section 5.3.1.1 bias of $\hat{Y}_{max}$ seems to have a limit value > 0 when the normalized curvature decreases to zero. The behaviour is also very different from bias of $\hat{Y}_{max}$ in unimodal regression described in Dahlbom (1986) where bias of the estimated maximum point increases to infinity when the normalized curvature increases. Therefore we want to examine bias of the estimated maximum point more carefully. This is first done by comparing $\hat{E}(\hat{Y}_{max})$ to the ordered $\hat{E}(Y_i)$, i=1,...,N, and then by examining the distinction between bias caused by the curvature and bias caused by the estimation method.

5.3.3.1. Comparison between $\hat{E}(\hat{Y}_{max})$ and the ordered $\hat{E}(Y_i)$.

A simulation study was made to examine the relation between y_{max} and the expectations of the ordered observations, $E(Y_i)$, i=1,....,N. Suppose we have a sample y_{ij}, j=1,...,$n(x_i)$, where $E(Y_i) = \mu(x_i)$ is a concave-unimodal function and $Y_i \sim N(\mu(x_i); \sigma(x_i))$. Also suppose that x_i are equidistant with the distance Δ between two successive x-observations. The simulation study was performed for $\mu(x)= -cx^2$ in the variation interval $0.05 \leq c \leq 3.00$ and for $-10 \leq x \leq 10$.

Denote the expected value of the largest observation $Y_{(N)}$ by $E_{(N)}$, the expected value of the second largest observation $Y_{(N-1)}$ by $E_{(N-1)}$ and so on. The result from 25 000 replications for $\Delta=1$, $n(x_i)=n=1$ and $\sigma(x_i)=1$ is shown in table 5.9.

TABLE 5.9.

NC	$\hat{E}(\hat{Y}_{max})$	$\hat{E}_{(N)}$	$\hat{E}_{(N-1)}$	$\hat{E}_{(N-2)}$	$\hat{E}_{(N-3)}$	$\hat{E}_{(N-4)}$
0.05	0.262	1.238	0.672	0.319	-0.212	-0.402
0.10	0.247	1.057	0.436	0.031	-0.321	-0.620
0.15	0.240	0.952	0.281	-0.158	-0.545	-0.929
0.20	0.267	0.856	0.158	-0.310	-0.749	-1.198
0.25	0.265	0.780	0.078	-0.449	-0.936	-1.499
0.30	0.263	0.722	0.022	-0.578	-1.106	-1.688
0.35	0.259	0.681	0.079	-0.682	-1.265	-1.910
0.40	0.256	0.658	0.150	-0.767	-1.421	-2.142
0.45	0.231	0.602	0.215	-0.828	-1.552	-2.346
0.50	0.222	0.561	0.260	-0.935	-1.718	-2.570
1.00	0.167	0.329	0.691	-1.610	-3.489	-4.576
1.50	0.099	0.183	1.088	-2.088	-5.433	-6.559
2.00	0.041	0.088	1.507	-2.555	-7.450	-8.578
2.50	-0.034	0.059	1.982	-3.037	-9.428	-10.566
3.00	-0.098	0.010	2.449	-3.564	-11.422	-12.570

Table 5.9. Random numbers were generated so that $Y \sim N(-cx^2;1)$ for 21 equidistant x-observations in the x-interval $-10 \leq x \leq 10$. The estimated expectation of $\hat{Y}_{max}$, $\hat{E}(\hat{Y}_{max})$, and the estimated expectations of the 5 largest ordered Y-observations, $\hat{E}_{(i)}$, i=N,N-1,...,N-4, were determined for $0.05 \leq c \leq 3.00$.

From the simulation study we see the relation between $\hat{E}(\hat{Y}_{max})$ and the estimated expected value of the ordered y-observations. For NC = 0.05 the y-value of the estimated maximum point lies between the fourth and the third biggest expected y-value. As NC increases $\hat{E}(\hat{Y}_{max})$ will get ever closer to the biggest expected value. The result can be summerized in table 5.10.

TABLE 5.10.	NC	$\hat{E}(\hat{Y}_{max})$ lies between
	0.05	$\hat{E}[Y_{(N-3)}]$ - $\hat{E}[Y_{(N-2)}]$
	0.10 - 0.15	$\hat{E}[Y_{(N-2)}]$ - $\hat{E}[Y_{(N-1)}]$
	0.20 - 3.00	$\hat{E}[Y_{(N-1)})$ - $\hat{E}(Y_{(N)}]$

Table 5.10. Random numbers were generated so that $Y \sim N(-cx^2;1)$ for 21 equidistant x-observations in the x-interval $-10 \leq x \leq 10$. The estimated expectation of $\hat{Y}_{max}$, $\hat{E}(\hat{Y}_{max})$, was estimated and compared to the estimated expectations of the largest ordered Y observations, $\hat{E}_{(i)}$, $i=N, N-1, \ldots, N-4$, for $0.05 \leq c \leq 3.00$.

These simulations indicate that $E(Y_{(N)})$ can serve as an upper limit of $E(\hat{Y}_{max})$. But $\hat{E}(Y_{(N)})$ will overestimate y_{max} more and more when $c \rightarrow 0$ since, for a regression function of infinite length,

$$\lim_{c \rightarrow 0} \hat{E}(Y_{(N)}) = \infty \quad \text{and} \quad \lim_{c \rightarrow 0} \hat{E}(\hat{Y}_{max}) = a,$$

where a is an unknown limit value. We have not been able to find a theoretical upper limit of $E(\hat{Y}_{max})$ that will hold for every value of the normalized curvature.

5.3.3.2. Distinction between bias caused by the curvature and bias caused by the estimation method.

When we use the proposed estimation method the following factors will have an influence on bias.

- The size of the curvature.
- The interval length between two successive x-observations.
- The number of x-observations in the examined interval.
- The variance of the observations.

First we will give general definitions of bias of the estimated maximum point. We can single out one part of the bias caused by the curvature and the fact that the position of the nearest x-value was uniformly distributed in the interval (-0.5Δ, 0.5Δ) centered at x_{max}. The other part is caused by the estimation method.

DEF:
$$\text{Bias}_1 = \int_{x_{max} - 0.5\Delta}^{x_{max} + 0.5\Delta} f(t) \frac{1}{\Delta} dt - y_{max}$$

is called bias of type 1 for the maximum point.

DEF:
$$\text{Bias}_2 = E(\hat{Y}_{max}) - \int_{x_{max} - 0.5\Delta}^{x_{max} + 0.5\Delta} f(t) \frac{1}{\Delta} dt$$

is called bias of type 2 for the maximum point.

This gives us the definition of the total bias:

$$Y_{bias}(x_{max}) = E(\hat{Y}_{max}) - y_{max}.$$

Bias caused by the curvature and the position of the nearest x-value compared to x_{max}:

This type of bias will always be non-positive for concave regression functions. If the position of the nearest x-value is uniformly distributed in the interval (-0.5Δ, 0.5Δ) centered at x_{max} the value of this type of bias is very easy to calculate in specific cases. For the regression function used in several simulation studies $\mu(x) = -cx^2$ we obtain for Δ=1, n=1 and σ(x)=1

$$\text{Bias}_1 = \int_{x_{max}-\frac{\Delta}{2}}^{x_{max}+\frac{\Delta}{2}} -cx^2 \frac{1}{\Delta} dx = -\frac{NC}{12}$$

In figure 5.3 we illustrate Bias_1 for Δ=1 and $0 \leq c \leq 16$.

FIGURE 5.3.

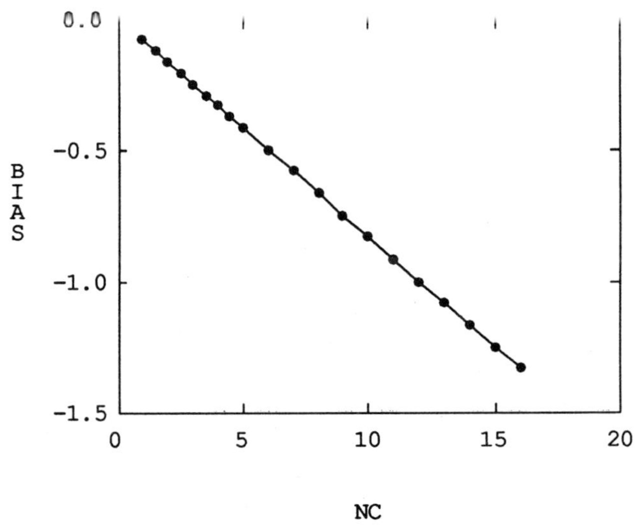

Fig. 5.3. Bias of type 1 is calculated for the regression function $\mu = -cx^2$.

Bias caused by the estimation method:

Bias of type 2, $Bias_2$, is difficult to examine by calculations since the estimation method is rather complicated. However if the normalized curvature, NC, is big enough the largest observation will also be the estimated maximum point. The largest observation will be obtained for x_1 or x_1-a, where x_1 is the nearest x-observation compared to x_{max}, where $x_1 \geq x_{max}$ and x_1-a is the corresponding x-value for which $x_1 - a \leq x_{max}$. In the specific situation, $y(x) = -NCx^2$, used in the simulations where the position of x_1 is uniformly randomized in the interval $[0, 0.5]$, the difference between $y(x_1)$ and $y(x_1-a)$ is obtained for a=1 as $NC(x_1-1)^2 - NCx_1^2 = NC(1-2x_1)$.

Denote $NC(1-2x_1)$ by v, where v is now uniformly distributed in the interval [0, NC]. Then bias of type 2 is obtained as

$$E[\max(y, y+v)] = \int_{v=0}^{NC} \int_{y=-\infty}^{\infty} y f_{max}(y) \frac{1}{NC} \, dy dv,$$

where $F_{max}(y) = \Phi(y)\Phi(y+v)$. Thus

$$E[\max(y, y+v)] = \int_{v=0}^{NC} \int_{y=-\infty}^{\infty} y\{\phi(y)\Phi(y+v) + \Phi(y)\phi(y+v)\} \frac{1}{NC} \, dy dv.$$

Since NC must be big we can approximately integrate the upper limit, NC, by ∞. Further

$$\int_{y=-\infty}^{\infty} \int_{v=0}^{\infty} y\phi(y)\Phi(y+v) dv dy = \int_{y=-\infty}^{\infty} \phi(y)(1-\Phi(y)) dy$$

and

$$\int_{y=-\infty}^{\infty} \int_{v=0}^{\infty} y\Phi(y)\phi(y+v) dv dy = \int_{y=-\infty}^{\infty} y\Phi(y)(1-\Phi(y)) dy$$

we obtain

$$E[\max(y,\ y+v)] \approx \frac{1}{NC} \int_{y=-\infty}^{\infty} \{\phi(y)(1-\Phi(y)) + y\Phi(y)(1-\Phi(y))\}dy.$$

But $\Phi(y)(1-\Phi(y))$ is even, which implies that the second expression is zero. Thus

$$E[\max(y,\ y+v)] \approx \frac{1}{NC}\left(\int_{y=-\infty}^{\infty} \phi(y)(1-\Phi(y))dy\right) = 1 - \int_{y=-\infty}^{\infty} \phi(y)\Phi(y)dy = 0.5/NC.$$

This means that in these situaitons $Bias_2 \approx 0.5/NC$.

FIGURE 5.4.

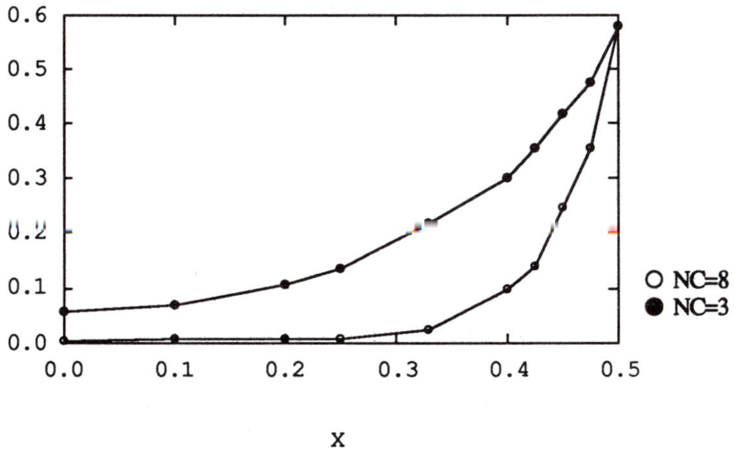

Fig. 5.4. Bias of type 2 for $\hat{Y}_{max}$ is estimated by using E(max(y, y+v)) on $y=-cx^2$ in the x-interval $0 \le x \le 0.5$.

A simulation study was made using 25 000 replications to estimate bias of type 2 for different values of NC and v. The results are shown in figure 5.4. Since in practice we don't know the value of v, bias of type 2 is obtained as the area under the curve in the interval $0 \leq x \leq 0.5$. In figure 5.4 we can see that when NC becomes bigger the corresponding curve approaches the x-axis and the straight line x=0.5.

Notice that this simulation study illustrates what happens when bias caused by the estimation method depends on the two x-observations closest to x_{max} only. Since this situation is a good approximation when NC is big enough, another simulation study was performed to examine the case when $Bias_2 \approx 1/2NC$. The simulation situation in section 5.3.1.1 for the x-interval $25 \leq x \leq 25$ and for the c-interval $0.20 \leq c \leq 16.0$, was used. The results from 5 000 replications are shown in table 5.11.

TABLE 5.11.

NC	$Bias_2$	NC	$Bias_2$	NC	$Bias_2$
0.20	0.289	2.50	0.189	8.0	0.060
0.30	0.317	3.00	0.168	9.0	0.056
0.40	0.325	3.50	0.148	10.0	0.050
0.50	0.331	4.00	0.130	11.0	0.045
0.75	0.322	4.50	0.116	12.0	0.042
1.00	0.290	5.00	0.100	13.0	0.038
1.50	0.249	6.00	0.086	14.0	0.036
2.00	0.216	7.00	0.072	16.0	0.031

Table 5.11. Random numbers were generated so that $Y \sim N(-cx^2;1)$ for 51 equidistant x-observations in the x-interval $-25 \leq x \leq 25$. Bias of type 2 for $\hat{Y}_{max}$ was estimated for $0.20 \leq c \leq 16.00$.

In this table we can see that for NC=6.0 we obtain $Bias_2 = 0.086$. When we calculate the fraction 1/2NC for NC=6.0, we obtain 0.083. Thus NC=6.0 is big enough for this approximation. From the table we can also see that bias of type 2 gives us the specific shape of increasing bias up to a maximum and then decreasing asymptotically to zero. This is also illustrated in figure 5.5.

FIGURE 5.5.

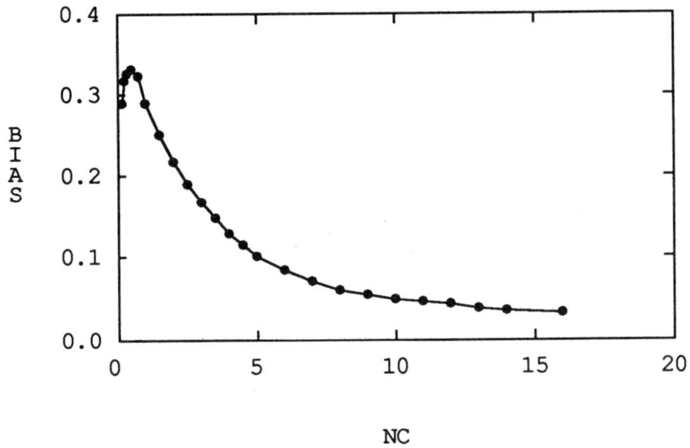

Fig. 5.5. Bias of type 2 for $\hat{Y}_{max}$ is estimated using concave regression on $\mu=-cx^2$ in the x-interval $-25 \le x \le 25$.

5.4.. Bias of $\hat{X}_{max}$ for concave-unimodal regression functions.

Since bias is dependent on the symmetry of the regression function we distinguish between symmetric and skew regression functions.

5.4.1. Bias of $\hat{X}_{max}$ for symmetric regression functions.

Suppose we have a situation that is totally symmetric with respect to x_{max}, ie when the distribution of Y, given X=x, is the same for $x=x_{max} + a$ and $x=x_{max} - a$. Then $\hat{X}_{max}$ is an unbiased estimator of x_{max}. An example of such a situation is used in the preceding simulation studies namely when the x-values are chosen equidistant and the position of the nearest x-value was uniformly distributed in the interval (-0.5Δ, 0.5Δ) centered at x_{max}. A detailed proof is given below.

THEOREM 5:1 Suppose that the distribution of Y, given X=x, is the same for $x=x_{max}+a$ and $x=x_{max}-a$. Also suppose that we observe at x points which are Δ equidistant with the position of the center uniformly distributed in [-0.5Δ, 0.5Δ]. Then $\hat{X}_{max}$ obtained by the concave-unimodal regression estimation method is an unbiased estimator of x_{max} .

PROOF: We want to show that for any $\delta > 0$

$$P(\hat{X}_{max} - x_{max} > \delta) = P(\hat{X}_{max} - x_{max} < -\delta).$$

To this purpose we introduce an artificial problem in the following way. Let $x_k^* = 2\,x_{max} - x_k$ and $Y_k^* = Y_k$. The problem with Y_k^* observed at x_k^*, k=1,2,...,N, now has the same distribution as the original problem of observing Y_k at x_k, k=1,2,...,N. Thus for any $\delta > 0$ $P(\hat{X}_{max} - x_{max} > \delta) = P(\hat{X}^*_{max} - x_{max} > \delta)$. But due to the symmetry of the estimation procedure $\hat{X}^*_{max} = 2x_{max} - \hat{X}_{max}$, ie

$$P(\hat{X}_{max} - x_{max} > \delta) = P(\hat{X}^*_{max} - x_{max} > \delta) =$$

$$P(2x_{max} - \hat{X}_{max} - x_{max} > \delta) = P(\hat{X}_{max} - x_{max} < -\delta).$$

Q.E.D.

5.4.2. The influence on bias of $\hat{X}_{max}$ by the skewness of the regression function.

In the preceding section we showed that $\hat{X}_{max}$ is an unbiased estimator of x_{max} for symmetric regression functions and symmetric positioning of x_i, i=1,...,N. When these two conditions are not fulfilled we can expect $\hat{X}_{max}$ to be biased. Therefore two situations have been examined to show the influence on bias, namely when the curvature is constant and when the regression function consists of piecewise linear functions.

5.4.2.1. The influence on bias of $\hat{X}_{max}$ by the skewness when the regression function is a third-degree polynomial.

A simulation study was performed to examine how skewness of the regression function influences bias. The regression function $\mu(x) = -cx^2 + dx^3$ was used for different values of c and d. The simulations were made for $-10 \le x \le 10$, $\Delta = 1$ and $\sigma = 1$. The position of the nearest x-value was uniformly distributed in the interval $(-0.5\Delta, 0.5\Delta)$ centered at x_{max}. The obtained results from 25 000 replications are shown in table 5.12.

The choice of the variation intervals for c and d was dependent on the fact that $\mu(x)$ is a third-degree function, which has a minimum in $x = 2c/3d$. For observations outside the examined x-interval [-10, 10] we have no longer a unimodal function. The chosen values of c and d give us the local minimum to the right of the chosen x-interval. The standard deviations of the means in the simulations lie between 0.04 - 0.17.

The results show that the examined values of d are too small compared to the curvature to show any pattern in the influence on bias of $\hat{X}_{max}$ by the skewness of the regression function. Thus a small skewness in the regression function will not give any essential bias of $\hat{X}_{max}$.

TABLE 5.12.

c	d = 0.000	d = 0.025	d = 0.050
0.8	0.000	0.010	0.000
0.9	0.000	0.009	0.002
1.0	0.000	0.012	0.016
1.5	0.000	0.006	0.010
2.0	0.000	0.003	0.006
2.5	0.000	0.003	0.005
3.0	0.000	0.002	0.003

Table 5.12. Random numbers were generated so that $Y \sim N(\mu(x);1)$ for 21 equidistant x-observations in the x-interval $-10 \leq x \leq 10$, where $\mu(x)=-cx^2+dx^3$. Bias of $\hat{X}_{max}$ was estimated for $0.8 \leq c \leq 3.0$ and $0.00 \leq d \leq 0.05$.

5.4.2.2. The influence on bias of $\hat{X}_{max}$ by the skewness for regression function with piecewise linear functions.

An analogous simulation study was performed using the regression function

$$\mu(x) = \begin{cases} cx & \text{for } x \leq 0 \\ -dx & \text{for } x > 0 \end{cases}$$

where $Y_i = \mu(x_i) + \varepsilon_i$ and $\varepsilon_i \sim N(0;1)$ in the interval $-25 \leq x \leq 25$. The values of the constants were d = 0.5, 1.0 and 1.5 while c varied from 0.5 to 3.0. In the same way as in the preceding study the x-observations were chosen equidistantly. The position of the nearest x-value was uniformly distributed in the interval $(-0.5\Delta, 0.5\Delta)$

centered at x_{max}. The simulation results from 25 000 replications are shown in table 5.13.

TABLE 5.13.

	d		
c	**0.50**	**1.00**	**1.50**
0.5	0.000	0.326	0.454
1.0	-0.320	0.000	0.142
1.5	-0.449	-0.133	0.000
2.0	-0.523	-0.217	-0.080
3.0	-0.599	-0.314	-0.183

Table 5.13. Random numbers were generated so that $Y \sim N(\mu(x);1)$ for 51 equidistant x-observations in the x-interval $-25 \leq x \leq 25$, where $\mu(x)=cx$ for $x \leq 0$ and $x=-dx$ for $x>0$. Bias of $\hat{X}_{max}$ was estimated for $0.5 \leq c \leq 3.0$ and $0.50 \leq d \leq 1.50$.

The standard errors of the means in the simulations lie between 0.002 - 0.006. From the simulation results we can see that bias of the estimated y_{max} is negative for all values of c and d under the diagonal where $c > d$ which means that the functions are negatively skew. On the upper side of the diagonal, where $d > c$, bias of the estimated y_{max} is positive. This means that the estimated x_{max} tends to the flat side of the regression function. From table 5.13 we can also see that for each value of d bias of the estimated y_{max} seems to decrease as c increases, and for each value of c, bias of the estimated y_{max} seems to increase when d increases.

5.5. Bias of the inflection point in sigmoid regression.

In sigmoid regression we will have an interesting problem namely the estimation of the inflection point (x_{inf}, y_{inf}). Since bias of both the estimated x_{inf} and the estimated y_{inf}, obtained by the estimation method, depends on symmetric conditions, both symmetric and non-symmetric regression functions are examined. For functions with certain symmetric conditions theoretical results are evaluated but for non-symmetric regression functions bias is examined by simulations.

5.5.1. Bias of the estimate of the inflection point of symmetric sigmoid functions.

Under certain symmetric conditions the estimation method gives us an unbiased estimate of the inflection point of the sigmoid regression function.

Suppose that we have a sample y_{ij}, $j=1,\ldots,n(x_i)$, where $E(Y_i) = \mu(x_i)$ is a sigmoid function and $Y_i \sim N(\mu(x_i); \sigma(x_i))$, $n(x_i)$ is the number of y-observations for x_i. The x-observations are chosen equidistant with the distance Δ between two successive x-observations. Denote the inflection point by (x_{inf}, y_{inf}). The position of the nearest x-value was uniformly distributed in the interval (-0.5Δ, 0.5Δ) centered at x_{inf}. Suppose that $\mu(x)$ satisfies $\mu(x) = 2\mu(x_{inf}) - \mu(2x_{inf}-x)$.

THEOREM 5:2 Under the conditions described above the estimation method will give us an unbiased estimate of the inflection point.

PROOF: Let x'_k and y'_k, k=1,...,N, denote the random variables

$$x'_k = 2x_{inf} - x_{N+1-k},$$
$$y'_k = 2y_{inf} - y_{N+1-k}.$$

Observe that x'_k and y'_k are artificial variables not possible to observe when x_{inf} and y_{inf} are unknown. The symmetric assumptions now imply that the problems with observations (x_k, y_k) k=1,...,N, and (x'_k, y'_k) k=1,...,N, follow the same model. Thus

$$P(\hat{X}_{inf} \le z) = P(\hat{X}'_{inf} \le z)$$
$$P(\hat{Y}_{inf} \le w) = P(\hat{Y}'_{inf} \le w).$$

In the estimation we fit the best sigmoid curve to data by least squares. This means here that if we could use sigmoid regression on the artificial variables x'_k and y'_k we would obtain

$$\hat{X}'_{inf} = 2x_{inf} - \hat{X}_{inf}$$
$$\hat{Y}'_{inf} = 2y_{inf} - \hat{Y}_{inf}.$$

Q.E.D.

COMMENT. If we have many observations in a big x-interval then it is not important to have different numbers of observations on each side of x_{inf}.

Examples of regression functions that fulfill these symmetric restrictions are

$$\mu(x) = \frac{e^x}{1 + e^x} \qquad x = -k\Delta, \ -(k-1)\Delta,...,0,......,k\Delta$$

$$\mu(x) = \Phi(x) \qquad x = -k\Delta,\ -(k-1)\Delta,\ldots,0,\ldots\ldots,k\Delta$$

and

$$\mu(x) = \begin{cases} 0.0 & \text{if } x \le -2.5 \\ 2x+5 & \text{if } -2.5 < x < 2.5 \\ 10.0 & \text{if } x \ge 2.5 \end{cases} \qquad x = -k\Delta,\ -(k-1)\Delta,\ldots,0,\ldots\ldots,k\Delta$$

5.5.2. Bias of the estimate of the inflection point of non-symmetric regression functions.

To be able to examine how bias is affected by a non-symmetric regression function we performed a simulation study. In this study we used N fixed equidistant x-values, where $\mu(x_i)$ was a non-symmetric sigmoid regression function having the inflection point in x=0. When we use the sigmoid estimation method the procedure will give us an inflection interval instead of an inflection point. The inflection point is estimated by the midpoint of the inflection interval.

The regression function examined was of the type $y_{ij} = \mu(x_i)+\varepsilon_{ij}$, $j=1,\ldots,n(x_i)$, where $\varepsilon_i \sim N(0, \sigma(x_i)^2)$, for bounded $\sigma(x_i)^2$. In the simulations we use $\mu_i(x) = 10*(1-e^{-e^x})$ in the x-interval $-10 \le x \le 10$. This regression function will have the inflection point $(x_{inf}, y_{inf}) = (0, 1- e^{-1})$. The distance between two successive x-observations, Δ, varied between 0.67 - 2.00. The position of the nearest x-value was uniformly distributed in the interval $(-0.5\Delta, 0.5\Delta)$ centered at x_{inf}. The result from 25 000 replications is rescaled and shown in table 5.14. From this table we can see that the estimation method will give us negatively biased estimates of both x_{inf} and y_{inf}. When the number of x-observations in a fixed x-interval increases both $\hat{Y}_{inf}$ and $\hat{X}_{inf}$ become less and less negatively biased. The results also showed that the estimated inflection points were tending to the flat side of the regression function. The standard errors of the means of $\hat{Y}_{inf}$ in the simulations lie between 0.010 - 0.012 and of $\hat{X}_{inf}$ between 0.003 - 0.005.

TABLE 5.14.

Number of x-observations N	Δ	Bias($\hat{Y}_{inf}$)	Bias($\hat{X}_{inf}$)
11	2.00	-0.995	-0.322
16	1.33	-0.896	-0.287
21	1.00	-0.842	-0.266
26	0.80	-0.783	-0.245
31	0.67	-0.759	-0.237

Table 5.14. Random numbers were generated so that $Y \sim N(\mu(x);1)$ for equidistant x-observations in the x-interval $-10 \leq x \leq 10$, where $\mu(x)$ was an extreme value function. Bias of $\hat{Y}_{inf}$ and bias of $\hat{X}_{inf}$ were estimated for $11 \leq N \leq 31$.

5.5.3. Comparison between bias of $\hat{X}_{inf}$ in sigmoid regression and bias of $\hat{X}_{max}$ using the corresponding slopes between successive observations.

A simulation study was performed to examine bias of the x-value of the estimated inflection point, $\hat{X}_{inf}$, using the extreme-value function $\mu(x)= 10*[1-\exp(-\exp(x))]$ in the interval $-10 \leq x \leq 10$. The aim of the study was to compare the results with the corresponding results for bias of $\hat{X}_{max}$ using the slopes between successive observations. In this study $Y= \mu(x) + \varepsilon$, where $\varepsilon \sim N(0, \sigma(x)^2)$. For each replication sigmoid regression was used. From the least squares sigmoid solution x_{inf} was estimated. For each replication we also calculated the slopes $(y_i-y_{i-1})/(x_i-x_{i-1})$, i=2,...,N, between successive observations. These slopes were used to estimate a unimodal curve from which we obtained $\hat{X}_{max}$. The x-observations were chosen equidistant with the distance Δ between two successive x-observations. The position of the nearest x-value was uniformly distributed in the interval (-0.5Δ, 0.5Δ) centered at $x_{inf} = x_{max}$. The number, N, of x-observations in the interval varied from 11 - 31. Of course we didn´t expect the results to be identical since we used two different estimation methods in the study. In the first estimation situation we used our proposed algorithm for sigmoid regression and in the second we used the algorithm for unimodal regression proposed by Frisén (1980). The results from the study using 25 000 replications are shown in table 5.15.

TABLE 5.15. **Bias**

N	Δ	$\hat{X}_{max}$	$\hat{X}_{inf}$
11	2.00	-0.284	-0.322
16	1.33	-0.249	-0.287
21	1.00	-0.255	-0.266
26	0.80	-0.267	-0.245
31	0.67	-0.275	-0.237

Table 5.15. Random numbers were generated so that Y ~ N(μ(x);1) for equidistant x-observations in the x-interval $-10 \leq x \leq 10$, where μ(x) was an extreme value function. Bias of $\hat{X}_{inf}$ using sigmoid regression was estimated for $11 \leq N \leq 31$. Bias of $\hat{X}_{max}$ using the slopes between successive observations was estimated for the same values of N.

From the table we can see that bias of both $\hat{X}_{inf}$ and $\hat{X}_{max}$ are negative for all values of N and that the size of bias of $\hat{X}_{inf}$ doesn't differ too much from the corresponding value of bias($\hat{X}_{max}$). However, bias of $\hat{X}_{inf}$ seems to decrease when N increases while bias of $\hat{X}_{max}$ rather seems to increase in the corresponding situations. The standard errors of the means of $\hat{X}_{inf}$ in the simulations lie between 0.003 - 0.005. The corresponding values of $\hat{X}_{max}$ are 0.006 - 0.007.

6. VARIANCE.

In this chapter we will examine some variance properties of the estimation method. This will be done both by theoretical calculations and by simulation studies. Since the piecewise linear regression functions contain both constructed and real observations we must distinguish between the variances for these two kind of points. We deal with the variance of the real observations by making some assumptions of the distribution of Y_k, k=1,..,N. But to get the variance of the constructed points we must perform theoretical calculations.

6.1. The variance of the constructed points for fixed bending points.

Suppose we have observations (x_k, y_k), k=1,..,N, with weights $w_k=1$ having the same variance $Var(Y_k) = \sigma^2$. Also suppose that we have M bending points with (m-1) x-observations between two successive bending points where $m \geq 2$. Thus we have M x-intervals, each containing m x-observations including the bending point. Let the x-observations be equidistant with the distance Δ between x_{i-1} and x_i, i=2,...,N. We can then determine the variances of the constructed points for $\Delta=1$ according the following:

Start in the first x-interval. Calculate the least squares estimate of a straight line, $f(x) = \bar{y}_1 + (x-\bar{x}_1)b_1$, for the observations, where $\bar{y}_1$ and $\bar{x}_1$ are the ordinary weighted averages. Use this straight line to obtain the first fiction point $(x_0^{(2)}, y_0^{(2)})$ with the weight $w_0^{(2)}$. The x-value of this constructed point is the bending point in the second x-interval. The variance, $(\tau_2)^2$, of this point is determined by:

$$(\tau_2)^2 = Var(y_0^{(2)}) = Var(\bar{y}_1 + (x_0^{(2)} - \bar{x}_1)b_1 = \frac{\sigma^2}{m} + \frac{(\frac{m+1}{2})^2 Var\, y_k w_k}{\sum_{k=0}^{m-1} (x_k - \bar{x}_1)^2 w_k} =$$

$$\sigma^2\left(\frac{1}{m} + (\frac{m+1}{2})^2 \frac{12}{m(m-1)(m+1)}\right) = \frac{\sigma^2}{w_0^{(2)}} .$$

Thus the fiction point $(x_0^{(2)}, y_0^{(2)})$ with the weight $w_0^{(2)}$ will have the variance $(\tau_2)^2 = \sigma^2/w_0^{(2)}$.

Now we continue to calculate the variance, $(\tau_3)^2$, of the fiction point $(x_0^{(3)}, y_0^{(3)})$ with the weight $w_0^{(3)}$. Determine the weighted least squares estimate of the regression line, $f(x) = \bar{y}_2 + (x-\bar{x}_2)b_2$, for the observations (x_i, y_i), i=m,....(2m-1). Notice that $x_m = x_0^{(2)}$ is the bending point in the second x-interval and $x_{2m} = x_0^{(3)}$ is the bending point in the third x-interval. Use the estimated regression line to obtain the fiction point $(x_0^{(3)}, y_0^{(3)})$ with the weight $w_0^{(3)}$. This variance, $(\tau_3)^2$, of this point is determined by the following:

$$(\tau_3)^2 = Var(y_0^{(3)}) = Var(\bar{y}_2 + (x_0^{(3)} - \bar{x}_2)b_2$$

where $$\bar{y}_2 = \frac{\sum_{k=m}^{2m-1} y_k w_k + y_0^{(2)} w_0^{(2)}}{m + w_0^{(2)}}$$

$$\bar{x}_2 = \frac{\sum_{k=0}^{m-1} (k+m) w_{k+m} + m w_0^{(2)}}{m + w_0^{(2)}}$$

Thus

$$\mathrm{Var}\left(\bar{y}_2 + (x_0^{(3)} - \bar{x}_2) b_2\right) =$$

$$\frac{\sigma^2}{m+w_0^{(2)}} \left(1 + \frac{(2m-\bar{x}_2)^2 (m+w_0^{(2)})}{\sum_{k=m}^{2m-1} (x_k - \bar{x}_2)^2 w_k + (m-\bar{x}_2)^2 w_0^{(2)}}\right) = \frac{\sigma^2}{w_0^{(3)}}.$$

Thus the variance of the fiction point $(x_0^{(3)}, y_0^{(3)})$ with the weight $w_0^{(3)}$ is $(\tau_3)^2 = \sigma^2/w_0^{(3)}$. More generally we can express the variance according to the following:

Determine the weighted least squares estimate of the regression line, $f(x) = \bar{y}_{i-1} + (x - \bar{x}_{i-1}) b_{i-1}$, for the observations in the (i-1):th x-interval including the fiction point $(x_0^{(i-1)}, y_0^{(i-1)})$ with the weight $w_0^{(i-1)}$. Use this estimated regression line to determine $(x_0^{(i)}, y_0^{(i)})$ with weight the $w_0^{(i)}$. The variance of the constructed point $(x_0^{(i)}, y_0^{(i)})$ with the weight $w_0^{(i)}$ is determined by:

$$(\tau_i)^2 = \mathrm{Var}(y_0^{(i)}) = \mathrm{Var}\left(\bar{y}_{i-1} + (x_0^{(i)} - \bar{x}_{i-1}) b_{i-1}\right) =$$

$$\frac{\sigma^2}{m+w_0^{(i-1)}} + \Big[(i-1)m - \bar{x}_{i-1})^2 *$$

$$\frac{\sigma^2}{\displaystyle\sum_{k=(i-2)m}^{(i-1)m-1}(x_k-\bar{x}_{i-1})^2 w_k + ((i-2)m-\bar{x}_{i-1})^2 w_0^{(i-1)}} \Bigg] = \frac{\sigma^2}{w_0^{(i)}} .$$

The calculations above give us the following theorem:

THEOREM 6:1 The successivly obtained variances, $(\tau_k)^2$, of the $y_0^{(k)}$-values, with the weights $w_0^{(k)}$, corresponding to the bending points, $x_0^{(k)}$ will converge to the limit value

$$\tau^2 = \frac{\sigma^2}{w_0} .$$

PROOF: In theorem 4:1 we proved that

$$\lim_{k\to\infty} w_0^{(k)} = w_0 .$$

From this relation follows that

$$\tau^2 = \lim_{k\to\infty} (\tau_k)^2 = \lim_{k\to\infty} \frac{\sigma^2}{w_0^{(k)}} = \frac{\sigma^2}{w_0} .$$

Q.E.D.

6.2. The influence on the variance of $\hat{Y}_i$ by the curvature of the increasing and concave regression function.

In the simulation studies described in chapter 5.2 concering bias we could see that the two separate situations with regression functions with constant curvature and regression functions consisting of linear functions were affected in somewhat different ways.

Therefore these two situations were examined to see if they would have any affect on the variances of the estimates. We were also interested to see if the estimates were improved by the fact that the curvature was constant.

6.2.1. Variance of $\hat{Y}(x_k)$ for increasing and concave regression functions with constant curvature.

A simulation study was performed to examine what variance properties the estimation method had for arbitrary points on the regression curve.

Suppose we have a sample $y_i = \mu(x_i) + \varepsilon_i$, i=1,....,N, where $\mu(x_i)$ is an increasing and concave regression function with equidistant x-observations in the interval $-d \leq x \leq 0$ and $\varepsilon_i \sim N(0; \sigma(x_i)^2)$, where $\sigma(x_i)^2$ are limited. The regression functions examined were of the type $\mu(x) = -cx^2$ for the variation interval $0.5 \leq c \leq 8.0$.

In the simulations with the regression function $\mu(x) = -cx^2$ the normalized curvature (NC) can be expressed in terms of the curvature (c), the distance between two successive x-observations (Δ), the variance (σ^2) and the number of y-observations (n) for each x_i by

$$NC = \frac{c\Delta^2}{\sigma/\sqrt{n}}$$

The simulation study was performed for

1) 15 equidistant x-observations in the interval $-7 \leq x \leq 0$, ie the distance between two successive x-observations was $\Delta=0.5$.
2) $n(x_i)=1$, i=1,...,15.
3) $\varepsilon_i \sim N(0; 1)$.

The result from 25 000 replications shows $Var[y(x_i)]/\sigma^2$, i=1,...,15, in table 6.1. Since the tendencies of the variances were the same for all examined values of the normalized curvature, we only illustrate the result for NC = 0.25, 0.50 and 0.75:

TABLE 6.1. $NC = \frac{c\Delta^2}{\sigma/\sqrt{n}}$

x_i	0.25	0.50	0.75
-7.0	0.783	0.891	0.915
-6.5	0.489	0.589	0.684
-6.0	0.399	0.517	0.664
-5.5	0.375	0.523	0.626
-5.0	0.377	0.527	0.615
-4.5	0.366	0.498	0.617
-4.0	0.390	0.530	0.620
-3.5	0.401	0.528	0.609
-3.0	0.394	0.536	0.642
-2.5	0.406	0.532	0.611
-2.0	0.406	0.547	0.624
-1.5	0.404	0.518	0.639
-1.0	0.409	0.547	0.663
-0.5	0.413	0.509	0.580
0.0	0.751	0.729	0.795

Table 6.1. Random numbers were generated so that $Y \sim N(-cx_i^2;1)$ for 15 equidistant x-observations in the x-interval $-7 \leq x \leq 0$. The variance of $\hat{Y}_i$ was determined for different values of the normalized curvature, NC.

From this table we see that the variance of $\hat{Y}_i$ compared to σ^2 is bigger at the end points of the regression curve. The other points seem to have the same variance with a slight decrease in the middle of the examined x-interval. The variance of $\hat{Y}_i$ also seem to increase when the normalized curvature increases.

6.2.2. Variance of $\hat{Y}(x_k)$ for increasing and concave regression functions consisting of linear functions.

Another simulation study was performed to examine how bias was affected by the fact that the regression functions consist of linear functions. To be able to compare the simulation results in this study to the results in the preceding section with constant curvature an analogous situation was examined.

Suppose we have a sample $y_k = \mu(x_k) + \varepsilon_k$, k=1,....,N, where $\mu(x_k)$ is an isotonic-concave regression function with equidistant x-observations in the interval $-d \leq x \leq 0$ and $\varepsilon_k \sim N(0; \sigma(x_k)^2)$ where $\sigma(x_k)^2$ are limited. The regression functions examined were of the type $\mu(x) = cx$ where the constant varied from 0.5 to 3.0.

In the simulations with the regression function $\mu(x) = cx$ the normalized slope, $N\beta$, can be expressed in terms of the constant (c), the distance between two successive x-observations (Δ), the variance ($\sigma(x)^2$) and the number of y-observations (n) for each x_k by

$$N\beta = \frac{c\Delta}{\sigma/\sqrt{n}}$$

The simulation study was performed for

1) 21 equidistant x-observations in the interval $-20 \leq x \leq 0$, ie the distance between two successive x-observations was Δ=1.0.
2) $n(x_k)=1$, k=1,...,21.
3) $\varepsilon_k \sim N(0, 1)$.

The result from 25 000 replications for $N\beta = 0.25$, 0.50 and 0.75 is shown in table 6.2. From this table we can see that we have the same tendency as for the situation with regression functions with constant curvature, ie the variances of $\hat{Y}_i$ compared to σ^2 are bigger

TABLE 6.2. $N\beta = \frac{c\Delta}{\sigma/\sqrt{n}}$

x_k	0.25	0.50	0.75
-20.0	0.441	0.440	0.439
-19.0	0.211	0.211	0.210
-18.0	0.159	0.159	0.159
-17.0	0.131	0.130	0.130
-16.0	0.111	0.110	0.110
-15.0	0.097	0.096	0.096
-14.0	0.086	0.086	0.086
-13.0	0.079	0.078	0.078
-12.0	0.073	0.073	0.073
-11.0	0.070	0.070	0.070
-10.0	0.069	0.069	0.069
-9.0	0.070	0.070	0.070
-8.0	0.073	0.073	0.073
-7.0	0.077	0.079	0.079
-6.0	0.083	0.086	0.086
-5.0	0.092	0.096	0.097
-4.0	0.104	0.109	0.110
-3.0	0.121	0.125	0.127
-2.0	0.146	0.150	0.152
-1.0	0.181	0.197	0.201
0.0	0.226	0.286	0.327

Table 6.2. Random numbers were generated so that $Y \sim N(\mu(x);1)$ for 21 equidistant x-observations in the x-interval $-20 \leq x \leq 0$, where $\mu(x)=cx$. The variances of $\hat{Y}_i$ were determined for different values of the normalized slope, $N\beta$.

at the end points of the regression curve. In the preceding simulation study the variances of $\hat{Y}_i$ compared to σ^2 also seemed to increase when the normalized curvature increased. When we have regression curves consisting of linear functions this is not the case for all x_i in the examined x-interval $-20 \leq x \leq 0$. Instead we obtained decreasing variances of $\hat{Y}_i$ compared to σ^2 for the lowest values of x_i, $-20 \leq x \leq -13$, when the normalized slope increased. In the middle of the x-interval, $-12 \leq x \leq -8$, we seemed to have constant variances for increasing values of the normalized slope. At the other end of the x-interval, $-7 \leq x \leq 0$, the variance of $\hat{Y}_i$ compared to σ^2 increased for increasing values of the normalized slope.

6.3. Some variance properties of concave-unimodal regression functions for stochastic X_i.

In addition to the estimation of $Y|x=x_i$ in concave-unimodal regression we also have the problem of estimating a maximum point. Theoretical calculations in this situation are too complicated and therefore this sector is examined by simulations.

6.3.1. The effect on the variance of $\hat{Y}_{max}$ by the curvature of the concave-unimodal regression function.

In the same way as for bias we are interested in what properties of the regression function that affect the variance of $\hat{Y}_{max}$. One of the properties we will examine is the influence on variance by the curvature. As for bias we can in this situation in a corresponding way distinguish two separate cases namely when the curvature is constant and when the regression functions consist of linear functions. Two simulation studies have been performed to see if the variances are affected in different ways by these cases. Since the maximum point of the regression function often is of greatest importance simulation studies were performed to examine the relation between $Var(Y_{max})$ and σ^2.

6.3.1.1. Variance of $\hat{Y}_{max}$ for concave-unimodal regression functions with constant curvature.

In the simulations 25 000 replications were used for the regression function $Y_i = \mu(x_i) + \varepsilon_i$, where $\mu(x) = -cx^2$ and $\varepsilon_i \sim N(0;1)$ for $-10 \leq x \leq 10$. The x-observations were chosen equidistantly. The position of the nearest x-value compared to x_{max} was uniformly distributed in the interval $(-0.5\Delta, 0.5\Delta)$ centered at x_{max}. The results were rescaled and shown in table 6.3.

TABLE 6.3.

$NC = \frac{c\Delta^2}{\sigma/\sqrt{n}}$	$\frac{\hat{V}ar(\hat{Y}_{max})}{\sigma^2}$	$NC = \frac{c\Delta^2}{\sigma/\sqrt{n}}$	$\frac{\hat{V}ar(\hat{Y}_{max})}{\sigma^2}$
0.05	0.207	1.50	0.683
0.10	0.271	1.75	0.717
0.20	0.352	2.00	0.746
0.50	0.485	2.25	0.773
0.75	0.549	2.50	0.793
1.00	0.602	2.75	0.818
1.25	0.645	3.00	0.830

Table 6.3. Random numbers were generated so that $Y \sim N(-cx_i^2;1)$ for 21 equidistant x-observations in the x-interval $-10 \leq x \leq 10$. The position of the nearest x-value was uniformly distributed in the interval $(-0.5\Delta, 0.5\Delta)$ centered at x_{max}. The variance of $\hat{Y}_{max}$ was determined for $0.05 \leq NC \leq 3.00$.

The conclusion from these results is that when NC increases $Var(\hat{Y}_{max})/\sigma^2$ increases too. The simulation results are also illustrated in figure 6.1.

FIGURE 6.1.

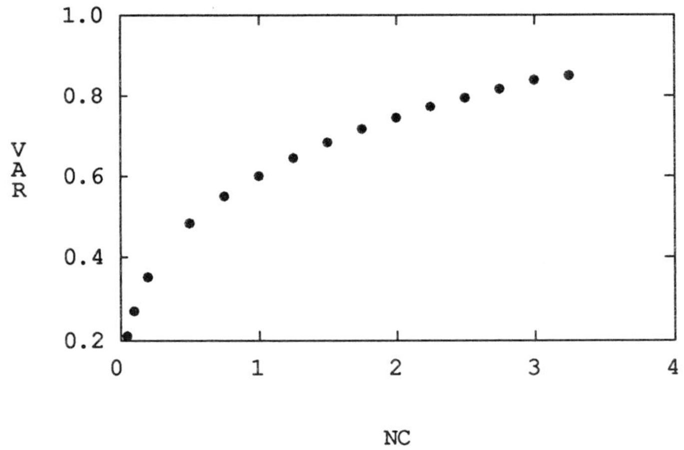

Fig 6.1. Random numbers were generated so that Y ~ $N(-cx_i^2;1)$ for 21 equidistant x-observations in the x-interval $-10 \leq x \leq 10$. The position of the nearest x-observation compared to x_{max} was randomized. $Var(\hat{Y}_{max})$ was estimated for different values of NC.

6.3.1.2. Variance of $\hat{Y}_{max}$ for concave-unimodal regression functions with piecewise linear functions.

To be able to compare this simulation study using regression functions consisting of piecewise linear functions with the preceding we performed a study as similar as possible. In this study we also used N fixed equidistant x-values, where $\mu(x_i)$ was a symmetric unimodal regression function with maximum in x = 0. The regression functions examined were of the type $y_{ij} = \mu(x) + \varepsilon_{ij}$, $j=1,....,n(x_i)$, where $\varepsilon_i \sim N(0, \sigma(x_i)^2)$, for limited $\sigma(x_i)^2$. In the simulations we use

$$\mu(x) = \begin{cases} cx & \text{for } x \leq 0 \\ -cx & \text{for } x > 0 \end{cases}$$

for the variation interval $0.50 \leq c \leq 2.00$. The corresponding x-interval was $-25 \leq x \leq 25$. The position of the nearest x-observation compared to x_{max} was randomized using a uniform distribution in the interval $[-0.5\Delta, 0.5\Delta]$. The result from 25 000 replications was rescaled and shown in table 6.4.

TABLE 6.4.	$N\beta = \dfrac{c\Delta}{\sigma/\sqrt{n}}$	$\dfrac{\hat{V}ar(\hat{Y}_{max})}{\sigma^2}$
	0.50	0.147
	0.75	0.177
	1.00	0.205
	1.25	0.232
	1.50	0.258
	1.75	0.283
	2.00	0.310

Table 6.4. Random numbers were generated so that $Y \sim N(\mu(x);1)$ for 51 equidistant x-observations in the x-interval $-25 \leq x \leq 25$, where $\mu(x)=cx$ for $x<0$ and $\mu(x)=-cx$ for $x \geq 0$. The variance of $\hat{Y}_{max}$ was estimated for different values of the normalized slope, $N\beta$.

From this table we can see that $\hat{V}ar(\hat{Y}_{max})/\sigma^2$ grows bigger when the normalized slope, $N\beta$, increases. The simulation results are also illustrated in figure 6.2 together with the results from the preceding simulation study using regression functions with constant curvature. The results from the preceding study are marked with black dots. From figure 6.2 we can see that the estimated $Var(\hat{Y}_{max})$ gets bigger when we have regression functions with constant curvature compared to regression functions with piecewise linear functions.

FIGURE 6.2.

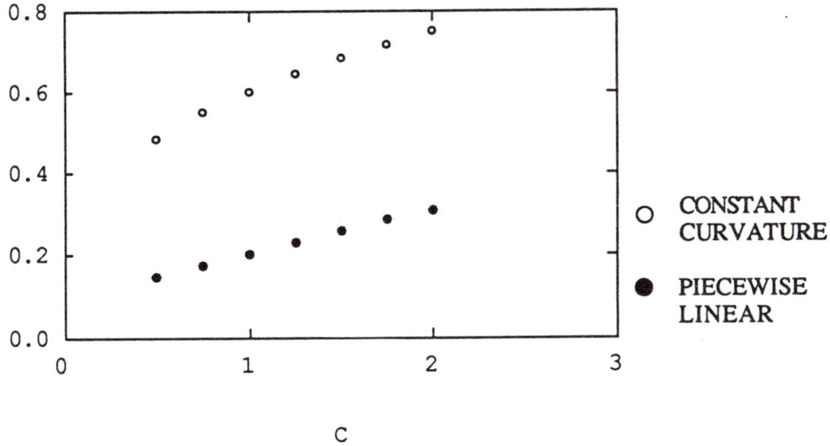

Fig 6.2. Random numbers were generated so that $Y \sim N(-cx_i;1)$ for 51 equidistant x-observations in the x-interval $-25 \leq x \leq 25$. The position of the nearest x-observation compared to x_{max} was randomized. $Var(\hat{Y}_{max})$ was estimated for different values of c. These estimates were compared to the estimated $Var(\hat{Y}_{max})$ obtained from the preceding simulation study for the same values of c.

6.3.2. The influence on the variance of $\hat{X}_{max}$ by the curvature of the concave-unimodal regression function.

Another result, that was obtained from the preceding simulation studies showed how the curvature of the regression function affected the variance of the estimated x_{max}. For $\hat{Y}_{max}$ we separated the two cases when the curvature of regression function was constant and when the regression function consists of piecewise linear functions. This was done for $\hat{X}_{max}$ too.

6.3.2.1. Variance of $\hat{X}_{max}$ for concave-unimodal regression functions with constant curvature.

In the simulation study 25 000 replications were used for the regression function $Y_i = \mu(x_i) + \varepsilon_i$, where $\mu(x) = -cx^2$ and $\varepsilon_i \sim N(0;1)$ for $-10 \leq x \leq 10$. The x-observations were chosen equidistantly with the distance Δ. The position of the nearest x-value compared to x_{max} was uniformly distributed in the interval $(-0.5\Delta, 0.5\Delta)$ centered at x_{max}. The results were rescaled and shown in table 6.5. The simulation results are also illustrated in figure 6.3.

TABLE 6.5.

$NC = \frac{c\Delta^2}{\sigma/\sqrt{n}}$	$\frac{\hat{V}ar(\hat{X}_{max})}{\Delta^2}$	$NC = \frac{c\Delta^2}{\sigma/\sqrt{n}}$	$\frac{\hat{V}ar(\hat{X}_{max})}{\Delta^2}$
0.05	4.598	1.50	0.244
0.10	2.484	1.75	0.213
0.20	1.411	2.00	0.190
0.50	0.634	2.25	0.173
0.75	0.443	2.50	0.159
1.00	0.347	2.75	0.147
1.25	0.285	3.00	0.138

Table 6.5. Random numbers were generated so that $Y \sim N(-cx_i^2;1)$ for 21 equidistant x-observations in the x-interval $-10 \leq x \leq 10$. The position of the nearest x-value was uniformly distributed in the interval $(-0.5\Delta, 0.5\Delta)$ centered at x_{max}. The variance of $\hat{X}_{max}$ was determined for $0.05 \leq NC \leq 3.00$.

This study shows that $Var(\hat{X}_{max})/\Delta^2$ decreases when NC increases. Thus the curvature has the opposite affect on $Var(\hat{X}_{max})/\Delta^2$ compared to $Var(\hat{Y}_{max})/\sigma^2$.

FIGURE 6.3

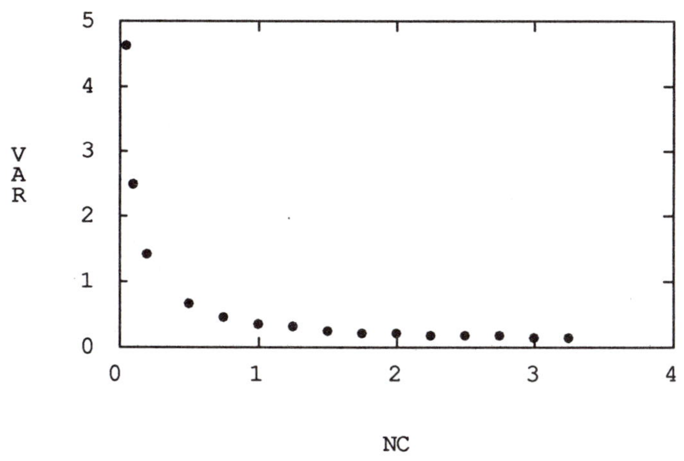

Fig 6.3. Random numbers were generated so that Y ~ $N(-cx_i^2;1)$ for 21 equidistant x-observations in the x-interval $-10 \leq x \leq 10$. The position of the nearest x-observation compared to x_{max} was randomized. $Var(x_{max})$ was estimated for different values of NC.

6.3.2.2. Variance of $\hat{X}_{max}$ for concave-unimodal regression functions with piecewise linear functions.

To be able to compare the results concering the variance of the estimated x_{max} in the two separate cases a simulation study, where the regression functions consisted of piecewise linear functions, was performed according to the following:

Since we wanted to compare this simulation study using piecewise linear regression functions with the results from the preceding study we performed another one that was as similar as possible. Therefore we used N fixed equidistant x-values, where $\mu(x_i)$ was a symmetric unimodal regression function with maximum in x = 0.

TABLE 6.6. $N\beta = \dfrac{c\Delta}{\sigma/\sqrt{n}}$	$\dfrac{\hat{V}ar(\hat{X}_{max})}{\Delta^2}$
0.50	0.873
0.75	0.481
1.00	0.318
1.25	0.234
1.50	0.186
1.75	0.159
2.00	0.140

Table 6.6. Random numbers were generated so that Y ~ N($\mu(x)$;1) for 51 equidistant x-observations in the x-interval $-25 \leq x \leq 25$, where $\mu(x)=cx$ for $x<0$ and $\mu(x)=-cx$ for $x\geq 0$. The variance of $\hat{X}_{max}$ was estimated for different values of the normalized slope, $N\beta$.

The regression functions examined were of the type $y_{ij} = \mu(x) + \varepsilon_{ij}$, $j=1,....,n(x_i)$, where $\varepsilon_i \sim N(0, \sigma(x_i)^2)$, for limited $\sigma(x_i)^2$. In the simulations we use

$$\mu(x) = \begin{cases} cx & \text{for } x \leq 0 \\ -cx & \text{for } x > 0 \end{cases}$$

for the variation interval $0.50 \leq c \leq 2.00$. The corresponding x-interval was $-25 \leq x \leq 25$. The position of the nearest x-observation compared to x_{max} was randomized using a uniform distribution in the interval $[-0.5\Delta, 0.5\Delta]$. The result from 25 000 replications was rescaled and shown in table 6.6. This study shows that $Var(\hat{X}_{max})/\Delta^2$ decreases when Nβ increases. The simulation results are also illustrated in figure 6.4 together with the corresponding results from the preceding simulation study.

FIGURE 6.4

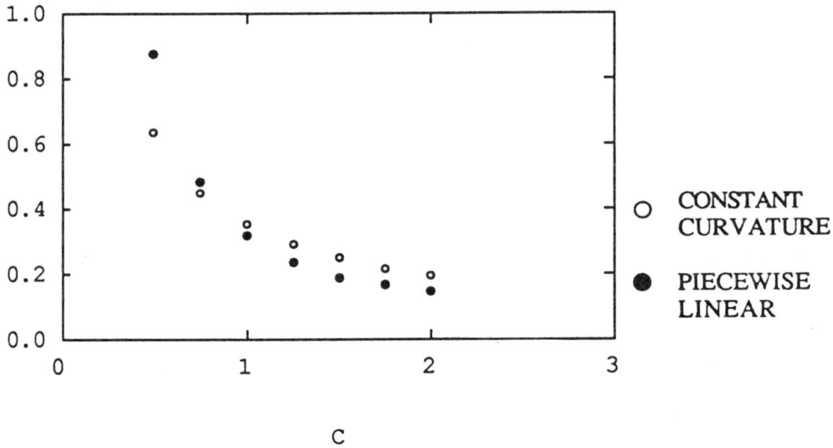

Fig 6.4. Random numbers were generated so that $Y \sim N(-cx_i;1)$ for 51 equidistant x-observations in the x-interval $-25 \leq x \leq 25$. The position of the nearest x-observation compared to x_{max} was randomized. $Var(x_{max})$ was estimated for different values of C. These estimates were compared to the estimated $Var(x_{max})$ obtained from the preceding simulation study for the same values of C.

The results from the preceding study are marked with black dots. From this figure we can see that the variances don´t differ too much. When we have a regression function consisting of piecewise linear functions $Var(\hat{X}_{max})/\Delta^2$ is bigger than the corresponding variance for constant curvature for small values of c but seems to be smaller when c increases.

6.3.3. The influence on the variance of some estimators by the skewness of the regression function.

Simulation studies were performed to see if skewness of the regression function influenced the variance of the examined regression estimators. The situations used in the simulations were $\mu(x) = -cx^2 + dx^3$ for $0.8 \leq c \leq 3.0$ and $0.00 \leq d \leq 0.05$ in the interval $-10 \leq x \leq 10$ and

$$\mu(x) = \begin{cases} cx & \text{for } x \leq 0 \\ -dx & \text{for } x > 0 \end{cases}$$

in the interval $-25 \leq x \leq 25$ for d = 0.5, 1.0 and 1.5 while c varied from 0.5 to 3.0. In both situations $Y_i = \mu(x_i) + \varepsilon_i$ and $\varepsilon_i \sim N(0;1)$. The x-observations were equidistant and the nearest x-value compared to x_{max} was uniformly distributed in the interval $(-0.5\Delta, 0.5\Delta)$. The result from 25 000 replications indicated that there was no influence by a small skewness on the variance of any of the regression estimators $Var(\hat{Y} \mid x_i)$, $Var(\hat{Y}_{max})$ and $Var(\hat{X}_{max})$.

6.4. Some variance properties of the inflection point of the sigmoid regression function for stochastic X_i.

In sigmoid regression the estimation of the inflection point (x_{inf}, y_{inf}) is of special interest. Since theoretical evaluations are very complicated in this situation the variances of both $\hat{X}_{inf}$ and $\hat{Y}_{inf}$, obtained by the estimation method, are examined by simulations.

6.4.1. The variance of the estimated y-value of the inflection point.

The variance of the estimated y-value of the inflection point was examined by simulations according to the following:

Suppose we have a sample $y_{ij} = \mu(x_i) + \varepsilon_{ij}$, j=1,....,$n(x_i)$, for each of N fixed equidistant x-values. Let $\mu(x) = E(Y|x)$ be a sigmoid regression function with the inflection point in x = 0 and $\varepsilon_i \sim N(0, \sigma(x_i)^2)$, where $\sigma(x_i)^2$ are limited. The simulation study was performed for the symmetric regression functions

$$\mu_1(x) = a + 10 * \frac{e^x}{1 + e^x}$$

$$\mu_2(x) = a + 10*\Phi(x)$$

and the non-symmetric regression function

$$\mu_3(x) = a + 10*(1-e^{-e^x}).$$

TABLE 6.7.

N	Δ	$\mu_1(x)$: $\frac{Var(\hat{Y}_{inf})}{\sigma^2}$	$\mu_2(x)$: $\frac{Var(\hat{Y}_{inf})}{\sigma^2}$	$\mu_3(x)$: $\frac{Var(\hat{Y}_{inf})}{\sigma^2}$
11	2.00	2.930	2.238	2.403
16	1.33	3.328	2.645	2.762
21	1.00	3.581	3.051	3.144
26	0.80	3.792	3.405	3.377
31	0.67	3.959	3.670	3.604

Table 6.7. Random numbers were generated so that $Y \sim N(\mu_i(x);1)$ for equidistant x-observations in the x-interval $-10 \le x \le 10$, where $\mu_1(x)$ was a logit function, $\mu_2(x)$ was a Gaussian function and $\mu_3(x)$ was an extreme-value function. The variances of $\hat{Y}_{inf}$ in these three cases were estimated for $11 \le N \le 31$.

FIGURE 6.5

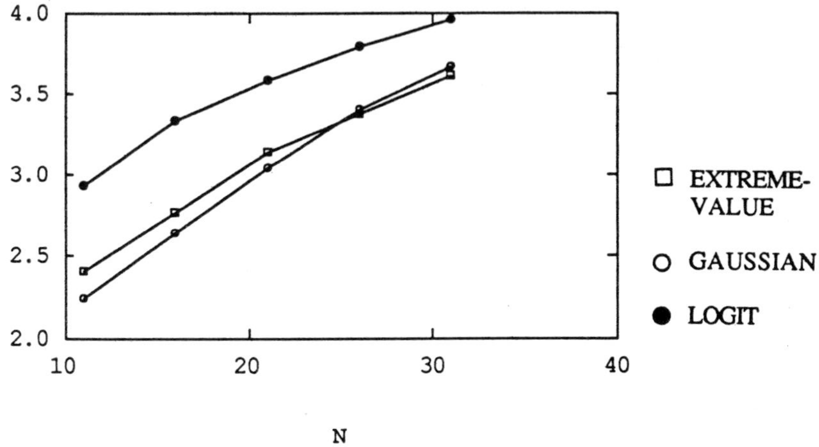

Fig 6.5. Random numbers were generated so that Y ~ N($-x_i$;1) for 11-31 equidistant x-observations in the x-interval $-10 \leq x \leq 10$. The position of the nearest x-observation compared to x_{inf} was randomized. Var(y_{inf}) was estimated for two symmetric regression functions and one non-symmetric.

In all situations we used equidistant x-observations in the interval $-10 \leq x \leq 10$. The number of x-observations, N, varied between 11-31 in the examined interval. The position of the nearest x-value compared to x_{inf} was uniformly distributed in the interval (-0.5Δ, 0.5Δ) centered at x_{inf}. In the simulations we also used $n(x_i)=1$ and $\sigma(x_i)=1$. In these situations the inflection point for the symmetric regression functions is known to be (0, 5) and for the non-symmetric function (0, $1-e^{-1}$). The variances of the estimated y_{inf} are obtained by 25 000 simulations. The result is shown in table 6.7. From this table we see that generally the variance increases when the number, N, of x-observations in the examined x-interval increases. The variances of the logit regression functions also seem to be slightly bigger than for the non-symmetric case. But the cumulative Gaussian regression functions seem to have the smallest variances except for N=26 and 31. The results from the simulation study are also shown in figure 6.5.

6.4.2. The variance of the estimated x-value of the inflection point.

Another result obtained from the preceding simulation study was the variance of the estimated x_{inf} obtained by the estimation method. In the simulation study we used the same three different regression functions as in the preceding simulation study to examine the difference between the variances obtained from the estimation method using symmetric regression functions compared

TABLE 6.8.

N	Δ	$\mu_1(x)$: $\frac{Var(\hat{X}_{inf})}{\Delta^2}$	$\mu_2(x)$: $\frac{Var(\hat{X}_{inf})}{\Delta^2}$	$\mu_3(x)$: $\frac{Var(\hat{X}_{inf})}{\Delta^2}$
11	2.00	0.294	0.121	0.153
16	1.33	0.565	0.211	0.267
21	1.00	0.940	0.361	0.453
26	0.80	1.418	0.559	0.681
31	0.67	2.016	0.893	1.075

Table 6.8. Random numbers were generated so that $Y \sim N(\mu_i(x);1)$ for equidistant x-observations in the x-interval $-10 \leq x \leq 10$, where $\mu_1(x)$ was a logit function, $\mu_2(x)$ was a Gaussian function and $\mu_3(x)$ was an extreme value function. The variances of $\hat{X}_{inf}$ in these three cases were estimated for $11 \leq N \leq 31$.

to non-symmetric. We also used the same assumptions, namely $Y = \mu(x) + \varepsilon$ where $\varepsilon \sim N(0;1)$ for $-10 \leq x \leq 10$. The x-observations were chosen equidistantly. The distance between two successive x-observations, Δ, varied between 0.67-2.00. The position of the nearest x-value compared to x_{inf} was uniformly distributed in the x-interval $(-0.5\Delta, 0.5\Delta)$ centered at x_{max}. The results from 25 000 replications are shown in the table 6.8.

FIGURE 6.6

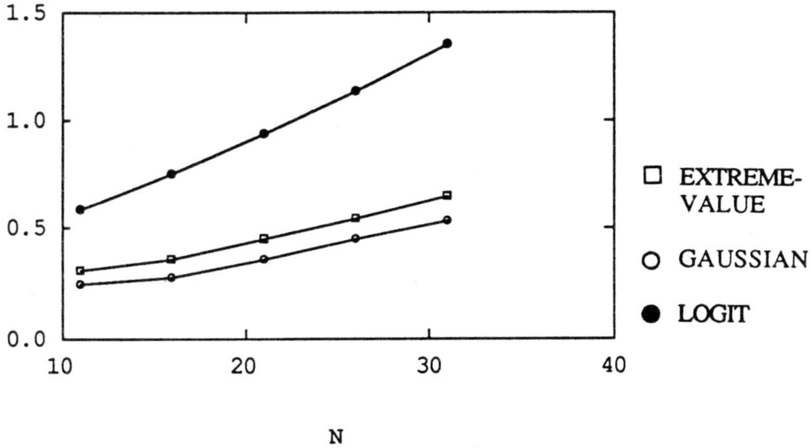

Fig 6.6. Random numbers were generated so that $Y \sim N(-x_i;1)$ for 11-31 equidistant x-observations in the x-interval $-10 \leq x \leq 10$. The position of the nearest x-observation compared to x_{inf} was randomized. $Var(X_{inf})$ was estimated for two symmetric regression functions and one non-symmetric.

From this table we see that $Var(\hat{X}_{max})/\Delta^2$ behaves in the same way as $Var(\hat{Y}_{max})/\sigma^2$, ie

1) the variances increase when the number, N, of x-observations in the examined x-interval increases.

2) the variances of the estimated x_{inf} using the logit regression function are slightly bigger than for the non-symmetric extreme-value function. The smallest variances are obtained for the cumulative Gaussian function. The results from the simulation study are also shown in figure 6.6.

6.4.3. Comparison between the variance of $\hat{X}_{inf}$ in sigmoid regression and the variance of $\hat{X}_{max}$ using the corresponding slopes between successive observations.

A simulation study was performed to examine the variance of $\hat{X}_{inf}$ using the extreme-value function $\mu(x) = 10*[1-\exp(-\exp(x))]$ in the interval $-10 \leq x \leq 10$. The aim of the study was to compare the results with the corresponding results for the variance of $\hat{X}_{max}$ using the slopes between successive observations. In this study $Y= \mu(x) + \varepsilon$, where $\varepsilon \sim N(0, \sigma(x)^2)$. For each replication sigmoid regression was used. From the least squares sigmoid solution x_{inf} was estimated. For each replication we also calculated the slopes $(y_i - y_{i-1})/(x_i - x_{i-1})$, i=2,...,N, between successive observations. These slopes were used to estimate a unimodal curve from which we obtained $\hat{X}_{max}$. The x-observations were chosen equidistantly with the distance Δ between two successive x-observations. The position of the nearest x-observation compared to $x_{inf} = x_{max}$ was uniformly randomized in the interval $[-0.5\Delta, 0.5\Delta]$ centered at x_{inf}. The number, N, of x-observations in the interval varied from 11 to 31.

Of course we didn't expect the results to be identical since we used two different estimation methods in the study. In the first estimation situation we used our proposed algorithm for sigmoid regression and in the second we used the algorithm for unimodel regression proposed by Frisén (1980). The results from the study using 25 000 replications are shown in table 6.9.

TABLE 6.9. **Variance**

N	Δ	$\hat{X}_{max}$	$\hat{X}_{inf}$
11	2.00	0.162	0.152
16	1.33	0.411	0.266
21	1.00	1.008	0.453
26	0.80	1.575	0.644
31	0.67	4.406	1.018

Table 6.9. Random numbers were generated so that $Y \sim N(\mu(x);1)$ for equidistant x-observations in the x-interval $-10 \leq x \leq 10$, where $\mu(x)$ was an extreme-value function. For each replication sigmoid regression was used and x_{inf} was estimated. For each replication we also calculated the slopes between successive observations. These slopes were used to estimate x_{max}. The variances of $\hat{X}_{inf}$ and $\hat{X}_{max}$ were obtained for $11 \leq N \leq 31$.

From the table we can see that the variances of both $\hat{X}_{inf}$ and $\hat{X}_{max}$ increase when N increases. The variances of $\hat{X}_{inf}$ are smaller than the corresponding value of $Var(\hat{X}_{max})$ for all values of N.

6.5. Estimation of $Var(\hat{Y})$.

In many situations we already have estimates of $Var(\hat{Y}_i)$ from another analogous survey. In other cases we can do some realistic theoretical assumptions. One assumption that is commonly used is that all y-observations, Y_i, i=1,..,N, have the same variance σ^2. If we don´t have any a priori information about the variance we can easily calculate an estimate if there is more than one y-observation for each x_i.

In some situations, however, we have only got one y-observation for each x_i. If $\mu(x_i)$ is a concave function and Y_i, j=1,...,N, are observations with $Y_i \sim N(\mu(x_i); \sigma^2(x_i))$ where the variance $\sigma^2(x_i)$ are assumed to be the same for all x_i, then the variance might be estimated by

$$(\hat{\sigma}_1)^2 = \frac{1}{N} \sum_{i=1}^{N} (y(x_i) - \hat{\mu}(x_i))^2 w_i \ .$$

When we use concave regression to estimate $\mu(x)$, $(\hat{\sigma}_1)^2$ will underestimate the variance σ^2. Therefore another estimation method has been developed for this situation.

Suppose that Y_i, i=1,...,N, have the same variance σ^2. We start by looking at three observations, since this is the smallest number of observations to form a concave (or convex) function.

Suppose that we have three observations, μ_1, μ_2, μ_3, forming a concave regression function. Let the corresponding observations on the least squares estimate of the regression line for these three observations be denoted by μ_1^*, μ_2^* and μ_3^*. Thus $\mu_1^* \geq \mu_1, \mu_2^* \leq \mu_2$ and $\mu_3^* \geq \mu_3$. Denote the distance between x_1 and x_2 by Δ_{12} and the distance between x_2 and x_3 by Δ_{23}. Then the corresponding distance between μ_1^* and μ_2^* will be $d_1=\hat{\beta}\Delta_{12}$ and between μ_2^* and μ_3^* $d_2=\hat{\beta}\Delta_{23}$ where $\hat{\beta}$ is the slope of the regression line through μ_1^*, μ_2^* and μ_3^*.

Case 1: If we have the special case that the concave regression function itself is a straight line we get equality for μ_i^* and μ_i, i=1,2,3, ie $\mu_1^* = \mu_1$, $\mu_2^* = \mu_2$ and $\mu_3^* = \mu_3$. If the x-observations are equidistant, ie if $\mu_2^* = \frac{1}{2}(\mu_1^* + \mu_3^*)$ then we can use the corresponding real observations Y_1, Y_2, Y_3 to calculate

$$s^2 = \frac{2}{3}\left[Y_2 - \frac{1}{2}(Y_1 + Y_3)\right]^2.$$

But usually this is not the case. Therefore we will use a more general formula.

The relation between μ_1^*, μ_2^* and μ_3^* can be written as

$$\mu_2^* = \frac{1}{(d_1+d_2)}\ (d_2\mu_1^* + d_1\mu_3^*).$$

Then we can do the corresponding calculation

$$s^2 = \frac{(d_1+d_2)^2}{(d_1+d_2)^2+d_1^2+d_2^2} * \left[Y_2 - \frac{1}{(d_1+d_2)}(d_2Y_1 + d_1Y_3)\right]^2.$$

Denote U= $Y_2 - \frac{1}{(d_1+d_2)}(d_2Y_1 + d_1Y_3)$. Since

$$E\left[U\right] = E\left[Y_2 - \frac{1}{(d_1+d_2)}(d_2Y_1 + d_1Y_3)\right] = \mu_2 - \frac{1}{(d_1+d_2)}(d_2\mu_1 + d_1\mu_3) = 0$$

we will obtain an ordinary $\chi^2(1)$ by $V^2 = \frac{s^2}{\sigma^2}$. From the relation $Var(U) = E(U^2) - [E(U)]^2$ we get $E(U^2) = Var(U) + [E(U)]^2$. Thus

$$E(U^2) = Var\left[Y_2 - \frac{1}{(d_1+d_2)}(d_2Y_1 + d_1Y_3)\right] - \left[E\left(Y_2 - \frac{1}{(d_1+d_2)}(d_2Y_1 + \right.\right.$$

$$d_1Y_3))\Big]^2 = \sigma^2 + \frac{d_2^{\,2}}{(d_1+d_2)^2}\sigma^2 + \frac{d_1^{\,2}}{(d_1+d_2)^2}\sigma^2 = \frac{(d_1+d_2)^2+d_1^{\,2}+d_2^{\,2}}{(d_1+d_2)^2}\,\sigma^2.$$

From this calculation we obtain that $E(s^2) = \sigma^2$ and therefore

$$s^2 = \frac{(d_1+d_2)^2}{(d_1+d_2)^2+d_1^{\,2}+d_2^{\,2}} * \Big[Y_2 - \frac{1}{(d_1+d_2)}(d_2Y_1 + d_1Y_3)\Big]^2$$

will serve as the variance estimate for these three observations. But usually a regression function consists of more than three observations. Then we can use

$$s^2 = \frac{\displaystyle\sum_{i=2}^{N-1} \frac{(d_{i-1}+d_i)^2}{(d_{i-1}+d_i)^2+d_{i-1}^{\,2}+d_i^{\,2}} * \Big[Y_i - \frac{1}{(d_{i-1}+d_i)}(d_iY_{i-1} + d_{i-1}Y_{i+1})\Big]^2}{N-2}$$

as an estimate of the variance, σ^2, of the observations.

Case 2: Suppose that the concave regression function consists of three observations, μ_1, μ_2 and μ_3. Also suppose that we don't have the special case where the concave regression function itself is a straight line. Then we obtain the unequalities $\mu_1^* > \mu_1$, $\mu_2^* < \mu_2$ and $\mu_3^* > \mu_3$. These order conditions give us

$$E\Big[Y_2 - \frac{1}{(d_1+d_2)}(d_2Y_1 + d_1Y_3)\Big] = \mu_2 - \frac{1}{(d_1+d_2)}(d_2\mu_1 + d_1\mu_3) > 0.$$

Thus $V^2 = \frac{s^2}{\sigma^2}$ is a non-central $\chi^2(1)$ with

$$V \sim N\left(\sqrt{\frac{(d_1+d_2)^2}{(d_1+d_2)^2+d_1^2+d_2^2}}\ \lambda \frac{1}{\sigma}\ ;\ 1\right)$$

where $\lambda = \mu_2 - \frac{1}{(d_1+d_2)}(d_2\mu_1 + d_1\mu_3)$. Since $\mathrm{Var}(V) = E(V^2) - [E(V)]^2$ we get

$$E(V^2) = \mathrm{Var}(V) + [E(V)]^2 = 1 + \frac{(d_1+d_2)^2}{(d_1+d_2)^2+d_1^2+d_2^2} * (\lambda \frac{1}{\sigma})^2 > 1.$$

From this unequality follows that

$$E(s^2) = E(\sigma^2 V^2) = \sigma^2 + \frac{(d_1+d_2)^2}{(d_1+d_2)^2+d_1^2+d_2^2} * (\lambda)^2 > \sigma^2.$$

Thus s^2 will overestimate the variance σ^2. If the regression function contains more that three observations the proposed estimate

$$s^2 = \frac{\sum_{i=2}^{N-1} \frac{(d_{i-1}+d_i)^2}{(d_{i-1}+d_i)^2+d_{i-1}^2+d_i^2} * \left[Y_i - \frac{1}{(d_{i-1}+d_i)}(d_i Y_{i-1} + d_{i-1} Y_{i+1})\right]^2}{N-2}$$

can be used as a rough estimate of the variance σ^2. To get a better estimate we can use

$$\sigma^2 = E(s^2) - \frac{(d_1+d_2)^2}{(d_1+d_2)^2+d_1^2+d_2^2} * (\lambda)^2$$

where λ is the average curvature of the regression function. This can be estimated according to the following:

Suppose we have least squares estimates of straight lines, $f(x_i) = \mu_i + \beta_i x_i$, $i=1,...,N-1$, through μ_i and μ_{i+1}. Thus the estimates consist of N-1 regression lines where the slopes $\beta_1 > \beta_2 >.....> \beta_{N-1}$ form a concave function. We can give an approximate model for the slope β_k for $\beta_1 > \beta_k > \beta_{N-1}$;

$$\beta_k = \beta_1 + (k-1)\frac{1}{N-1}(\beta_{N-1} - \beta_1)$$

From this model we obtain

$$\beta_k - \beta_{k-1} = \frac{\beta_{N-1} - \beta_1}{N-1}$$

For the first two x-intervals we get

$$\lambda = \mu_2 - \frac{1}{(d_1+d_2)}(d_2\mu_1 + d_1\mu_3) = \frac{1}{(d_1+d_2)}(d_2\beta_1 - d_1\beta_2) =$$

$$\frac{1}{(d_1+d_2)}\left[d_1\left(\frac{\beta_1 - \beta_{N-1}}{N-1}\right) + (d_2 - d_1)\beta_1\right].$$

This gives us the desired relation for the two first x-intervals

$$\frac{(d_1+d_2)^2}{(d_1+d_2)^2+d_1^2+d_2^2}*(\lambda)^2 = \frac{(d_1+d_2)^2}{(d_1+d_2)^2+d_1^2+d_2^2}*$$

$$\left[\frac{1}{(d_1+d_2)}\left[d_1\left(\frac{\beta_1 - \beta_{N-1}}{N-1}\right) + (d_2 - d_1)\beta_1\right]\right]^2.$$

For any two x-interval we get the corresponding expression

$$\frac{(d_{i-1}+d_i)^2}{(d_{i-1}+d_i)^2+d_{i-1}{}^2+d_i{}^2}*(\lambda)^2 = \frac{(d_{i-1}+d_i)^2}{(d_{i-1}+d_i)^2+d_{i-1}{}^2+d_i{}^2}*$$

$$\left[\frac{1}{(d_{i-1}+d_i)}\left[d_{i-1}\left(\frac{\beta_1 - \beta_{N-1}}{N-1}\right) + (d_i - d_{i-1})\beta_1\right]\right]^2.$$

Since in this expression we have β_1 and β_{N-1} we must estimate them. This can be done according to the following:

Suppose we have a least squares estimate of a concave regression function with M bending points $x_0^{(1)} < x_0^{(2)} <\ldots\ldots< x_0^{(M)}$. Thus the estimate consists of M regression lines where the slopes $\hat{\beta}_1 > \hat{\beta}_2 >\ldots>$ $\hat{\beta}_M$ form a concave function. The estimates of β_1 and β_{N-1} are $\hat{\beta}_1$ and $\hat{\beta}_M$, where $\hat{\beta}_1$ and $\hat{\beta}_M$ are the end slopes in the concave estimate of the whole regression function $\mu(x_i)$, i=1,...,N. The proposed variance estimate will be

$$s^2 = \frac{1}{N-2}\left[\sum_{i=2}^{N-1}\left\{\frac{(d_{i-1}+d_i)^2}{(d_{i-1}+d_i)^2+d_{i-1}{}^2+d_i{}^2}*\left(Y_i - \frac{1}{(d_{i-1}+d_i)}(d_iY_{i-1} +\right.\right.\right.$$

$$\left.\left.\left. d_{i-1}Y_{i+1})\right)^2 - \left(\frac{1}{(d_{i-1}+d_i)}\left[d_{i-1}\left(\frac{\hat{\beta}_1 - \hat{\beta}_M}{N-1}\right) + (d_i - d_{i-1})b_1\right]\right)^2\right\}\right].$$

If the x-observations are equidistant this formula will be simplified to

$$\hat{\sigma}^2 = \frac{\frac{2}{3}}{N-2} \sum_{i=2}^{N-1} \left[Y_i - \frac{1}{2}(Y_{i-1} + Y_{i+1}) \right]^2 - \frac{2}{3}\frac{1}{4}\left(\frac{\hat{\beta}_1 - \hat{\beta}_M}{N-1}\right)^2.$$

From the proposed formula we can also see that if we have the described situation in case 1, that the regression function already is a straight line, then we obtain $\beta_1 = \beta_{N-1}$ and the correction term vanishes. This gives us the variance estimate that was proposed in case 1.

6.5.1. Improvement of the proposed variance estimation method for regression functions with big curvature.

The estimation method proposed in the preceding section is based on short linear pieces, which are weighted together. If the regression function has a big curvature we can expect that the variance will be overestimated rather much. In this situation it might be appropriate to consider second-degree functions instead of linear pieces. If we use the assumption that all Y_i have the same variance then the corresponding variance estimation method can be evaluated according to the following discussion:

Since we will use second-degree functions three parameters are to be estimated. Therefore we start by looking at five observations forming a concave (or convex) function. This gives us two degrees of freedom for the variance estimate.

Case 1: Suppose we have equidistant x-observations with the distance Δ between two successive observations. Let the points be centered at x=0.0. Thus the five observations will be Y_{-2}, Y_{-1}, Y_0, Y_1 and Y_2 and the corresponding x-observations -2Δ, $-\Delta$, 0.0, Δ and 2Δ.

The parameters β_i, i=0,1,2, to be estimated are contained in the function y= $\beta_0 + \beta_1 x + \beta_2(x^2 - \gamma)$. We can simplify the evaluation if we calculate an appropriate value of γ to obtain orthogonality:

$$((-2\Delta)^2-\gamma)*1+((-\Delta)^2-\gamma)* \ -\gamma*1+(\Delta^2-\gamma)*1+((2\Delta)^2-\gamma)*1 = 0 \quad \Leftrightarrow \quad \gamma = 2\Delta^2.$$

Then we obtain the estimates

$$\hat{\beta}_0 = \frac{1}{5} \sum_{i=-2}^{2} Y_i$$

$$\hat{\beta}_1 = \frac{\sum_{i=-2}^{2} i*Y_i}{10\Delta}$$

$$\hat{\beta}_2 = \frac{2Y_{-2} - Y_{-1} - 2Y_0 - Y_1 + 2Y_2}{14\Delta^2}.$$

From the five points we obtain the following contribution to the variance estimate

$$\frac{1}{2} \sum_{i=-2}^{2} \left[Y_i - (\hat{\beta}_0 + \hat{\beta}_1 x_i + \hat{\beta}_2(x_i^2 - 2\Delta^2))\right]^2 .$$

This expression can be evaluated to hold for several second-degree pieces. Suppose that we obtain the sets of five observations successively. Denote the estimated parameters from the k:th set by $\hat{\beta}_i^{(k)}$, i=0,1,2. Thus we can use

$$S^2 = \frac{\sum_{k=3}^{N-2} \frac{1}{2} \sum_{i=k-2}^{k+2} \left[(Y_i - Y_k) - \{ \hat{\beta}_0^{(k)} + \hat{\beta}_1^{(k)}(x_i - x_k) + \hat{\beta}_2^{(k)}((x_i - x_k)^2 - 2\Delta^2) \} \right]^2}{N-4}$$

as an estimate of the variance, σ^2, of the observations on the whole curve.

Case 2: If the x-observations are not equidistant then it is no simplification to central the x-observations or determine a value of γ to obtain orthogonality. Instead we will use the ordinary normal equations to estimate β_i, i=0,1,2:

$$\sum_{i=1}^{N} Y_i = N\hat{\beta}_0 + \hat{\beta}_1 \sum_{i=1}^{N} x_i + \hat{\beta}_2 \sum_{i=1}^{N} x_i^2$$

$$\sum_{i=1}^{N} x_i Y_i = \hat{\beta}_0 \sum_{i=1}^{N} x_i + \hat{\beta}_1 \sum_{i=1}^{N} x_i^2 + \hat{\beta}_2 \sum_{i=1}^{N} x_i^3$$

$$\sum_{i=1}^{N} x_i^2 Y_i = \hat{\beta}_0 \sum_{i=1}^{N} x_i^2 + \hat{\beta}_1 \sum_{i=1}^{N} x_i^3 + \hat{\beta}_2 \sum_{i=1}^{N} x_i^4.$$

Thus the corresponding estimate of the variance, σ^2, for the observations on the whole curve is reduced to

$$S^2 = \frac{\sum_{k=3}^{N-2} \frac{1}{2} \sum_{i=k-2}^{k+2} \left[Y_i - \{ \hat{\beta}_0^{(k)} + \hat{\beta}_1^{(k)} x_i + \hat{\beta}_2^{(k)} x_i^2 \} \right]^2}{N-4}.$$

If the curvature of the regression function is big this alternative will not overestimate the variance as much as the one proposed for linear pieces. However, for functions with small curvature we will obtain almost the same size of the estimate of the variance. In practice this is probably a minor problem. Therefore the simulation studies in the following sections are made for the estimation method proposed for linear pieces.

6.5.2. Properties of the proposed variance estimation method for a regression function with constant curvature.

A simulation study was performed to examine the properties of the proposed variance estimation method for a regression function with constant curvature.

We used 25 000 replications for the regression function $Y_i = \mu(x_i) + \varepsilon_i$, where $\mu(x) = -cx^2$ and $\varepsilon_i \sim N(0;1)$ for $-10 \leq x \leq 10$.

TABLE 6.10.

$NC = \dfrac{c\Delta^2}{\sigma/\sqrt{n}}$	$\hat{E}(s^2)$	corr	$\hat{\sigma}^2$	$\hat{\sigma}$
0.0125	1.0010	0.0009	1.0001	1.000
0.05	1.0068	0.0036	1.0032	1.000
0.20	1.0258	0.0184	1.0074	1.004
0.50	1.1657	0.1024	1.0633	1.031
0.75	1.3740	0.2356	1.1377	1.067
1.00	1.6655	0.4421	1.2222	1.106
1.50	2.4987	1.0994	1.3977	1.182
2.00	3.6568	2.1300	1.5335	1.238

Table 6.10. Random numbers were generated so that $Y \sim N(-cx^2;1)$ for equidistant x-observations in the x-interval $-10 \leq x \leq 10$. The variances, $\hat{\sigma}^2 = \hat{E}(s^2)$ - corr, were estimated according to the proposed estimation method for $0.0125 \leq NC \leq 2.00$.

The x-observations were chosen equidistantly. The position of the nearest x-value compared to x_{max} was uniformly distributed in the interval (-0.5Δ, 0.5Δ). In the simulations both $\hat{E}(s^2)$, the estimated correction term, $\hat{\sigma}^2$ and the corresponding variations of the estimates were determined. The results were rescaled and shown in table 6.10. The estimated correction term is denoted by corr.
From this table we can see that both $\hat{E}(s^2)$, the estimated correction term and $\hat{\sigma}^2$ increase when NC increases. But still when NC=2.0, $\hat{\sigma}^2$ will be rather moderate. Bigger values of NC are useless to examine since then the observations almost themselves will form a concave regression function. From the table we can see that $\hat{\sigma}^2$ seems to converge to σ^2 when the normalized curvature decreases to zero.

FIGURE 6.7.

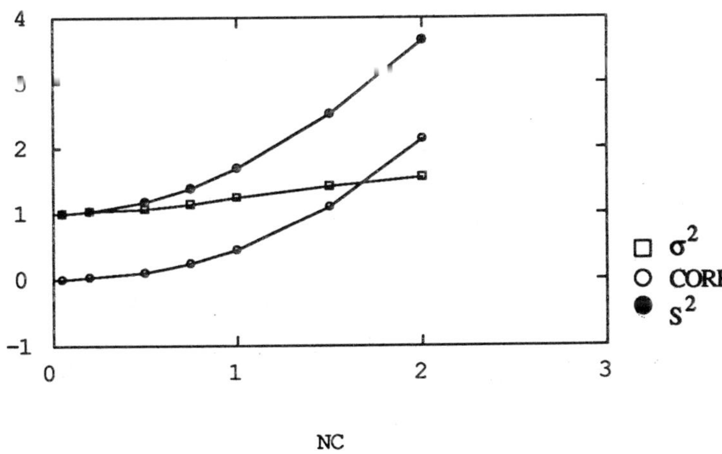

Fig 6.7. Random numbers were generated so that $Y \sim N(-cx_i^2;1)$ for 21 equidistant x-observations in the x-interval $-10 \le x \le 10$. The position of the nearest x-observation compared to x_{max} was randomized. $\hat{E}(s^2)$, corr and $\hat{\sigma}^2$ were estimated for different values of NC.

The simulation results also indicate that $Var(\hat{\sigma}^2)$ increases when NC increases. The results concerning $\hat{E}(s^2)$, the estimated correction term and $\hat{\sigma}^2$ are illustrated in the figures 6.7.

To examine the variation in the results the 3:rd, μ_3, and the 4:th, μ_4, moments were determined and the estimates of $\mu_3/(\text{stand dev}(\sigma)^3)$ and $\mu_4/(\text{stand dev}(\sigma)^4)$ were calculated for different values of the normalized curvature, NC, together with the estimated distributions. The results were rescaled and shown in table 6.11.

TABLE 6.11.	NC	$\frac{\mu_3}{(\text{stand dev}(\sigma^3))}$	$\frac{\mu_4}{(\text{stand dev}(\sigma^4))}$
	0.05	1.008	4.495
	0.20	1.008	4.498
	0.50	0.981	4.424
	0.75	0.900	4.245
	1.00	0.801	3.982
	1.50	0.707	3.217
	2.00	1.197	3.973

Table 6.11. Random numbers were generated so that $Y \sim N(-cx^2;1)$ for equidistant x-observations in the x-interval $-10 \le x \le 10$. The standardized skewness and the standardized kurtosis were obtained for the variance estimates for $0.05 \le NC \le 2.00$.

From this table we can see that the standardized skewness was around 1.0 for all values of NC and the standardized kurtosis near 4.0. To see if the estimate could be approximated by a χ^2-distribution, we also more carefully examined the estimated distribution of $\hat{\sigma}^2$. This knowledge can be very useful if we try to determine approximately 95% limits of the estimate. The estimated cumulative distribution for NC=0.5 obtained from the simulations is shown in table 6.12.

TABLE 6.12.

$\hat{\sigma}^2$	$F(\hat{\sigma}^2)$	$\hat{\sigma}^2$	$F(\hat{\sigma}^2)$	$\hat{\sigma}^2$	$F(\hat{\sigma}^2)$	$\hat{\sigma}^2$	$F(\hat{\sigma}^2)$
0.250	0.0001	1.150	0.6299	2.050	0.9665	2.950	0.9980
0.275	0.0016	1.175	0.6501	2.075	0.9691	2.975	0.9982
0.300	0.0042	1.200	0.6696	2.100	0.9715	3.000	0.9984
0.325	0.0079	1.225	0.6874	2.125	0.9736	3.025	0.9986
0.350	0.0124	1.250	0.7037	2.150	0.9756	3.050	0.9988
0.375	0.0178	1.275	0.7218	2.175	0.9774	3.075	0.9989
0.400	0.0247	1.300	0.7365	2.200	0.9790	3.100	0.9990
0.425	0.0319	1.325	0.7516	2.225	0.9805	3.125	0.9991
0.450	0.0406	1.350	0.7665	2.250	0.9822	3.150	0.9992
0.475	0.0497	1.375	0.7813	2.275	0.9836	3.175	0.9992
0.500	0.0605	1.400	0.7946	2.300	0.9849	3.200	0.9993
0.525	0.0730	1.425	0.8076	2.325	0.9864	3.225	0.9993
0.550	0.0868	1.450	0.8189	2.350	0.9876	3.250	0.9994
0.575	0.1041	1.475	0.8302	2.375	0.9889	3.275	0.9994
0.600	0.1208	1.500	0.8396	2.400	0.9899	3.300	0.9995
0.625	0.1408	1.525	0.8486	2.425	0.9905	3.325	0.9996
0.650	0.1607	1.550	0.8580	2.450	0.9914	3.350	0.9996
0.675	0.1816	1.575	0.8681	2.475	0.9919	3.375	0.9997
0.700	0.2024	1.600	0.8768	2.500	0.9924	3.400	0.9997
0.725	0.2264	1.625	0.8840	2.525	0.9927	3.425	0.9998
0.750	0.2523	1.650	0.8684	2.550	0.9932	3.450	0.9998
0.775	0.2767	1.675	0.8985	2.575	0.9938	3.475	0.9998
0.800	0.3001	1.700	0.9051	2.600	0.9941	3.500	0.9999
0.825	0.3254	1.725	0.9112	2.625	0.9948	3.525	0.9999
0.850	0.3504	1.750	0.9170	2.650	0.9951	3.550	0.9999
0.875	0.3751	1.775	0.9221	2.675	0.9955	3.575	0.9999
0.900	0.3997	1.800	0.9279	2.700	0.9960	3.600	0.9999
0.925	0.4248	1.825	0.9332	2.725	0.9963	3.625	0.9999
0.950	0.4485	1.850	0.9386	2.750	0.9968	3.650	0.9999
0.975	0.4727	1.875	0.9434	2.775	0.9970	3.675	0.9999
1.000	0.4956	1.900	0.9473	2.800	0.9972	3.700	0.9999
1.025	0.5174	1.925	0.9515	2.825	0.9974	3.725	0.9999
1.050	0.5410	1.950	0.9544	2.850	0.9976	3.750	0.9999
1.075	0.5631	1.975	0.9574	2.875	0.9978	3.775	0.9999
1.100	0.5866	2.000	0.9604	2.900	0.9978	3.800	0.9999
1.125	0.6086	2.025	0.9628	2.925	0.9980	3.825	1.0000

Table 6.12. Random numbers were generated so that $Y \sim N(-cx^2;1)$ for equidistant x-observations in the x-interval $-10 \leq x \leq 10$. The distribution of $\hat{\sigma}^2$, using the proposed estimation method, was estimated for NC = 0.50.

In figure 6.8 we can see the shapes of the estimated probability distributions of the estimated variance for NC=0.05, 0.5 and 2.0.

FIGURE 6.8.

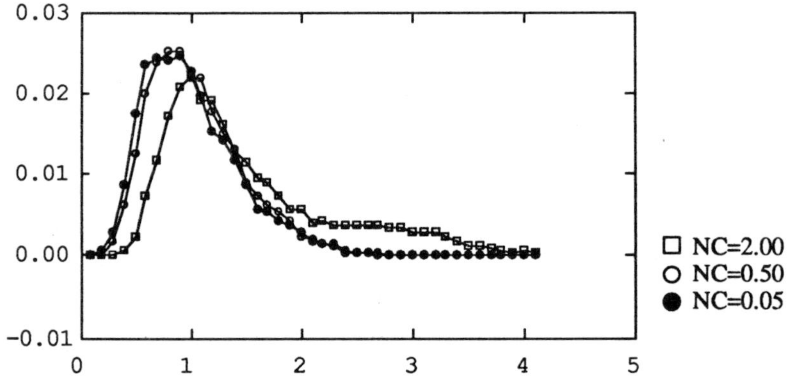

$\hat{\sigma}^2$

Fig 6.8. Random numbers were generated so that Y ~ $N(-cx_i^2;1)$ for 21 equidistant x-observations in the x-interval $-10 \leq x \leq 10$. The position of the nearest x-observation compared to x_{max} was randomized. The estimated distribution of $\hat{\sigma}^2$ were determined for NC =0.05, 0.5 and 2.0.

In this figure we can see that the estimated distributions look something like a χ^2-distribution. From the same simulation study we also obtained the 0.025-limits and the 0.975-limits of the estimated distributions for $0.20 \leq$ NC ≤ 2.00. The results are shown in table 6.13. This table shows that both the value of the lower and the upper limit increase when NC increases.

TABLE 6.13.

NC	0.20	0.50	0.75	1.00	1.50	2.00
lower limit	0.350	0.400	0.425	0.450	0.505	0.600
upper limit	2.075	2.150	2.225	2.350	2.765	3.450

Table 6.13. Random numbers were generated so that $Y \sim N(-cx^2;1)$ for equidistant x-observations in the x-interval $-10 \le x \le 10$. The 0.025 and 0.975 limits of the estimated cumulative distribution of $\hat{\sigma}^2$, using the proposed estimation method, were obtained for $0.20 \le NC \le 2.00$.

FIGURE 6.9.

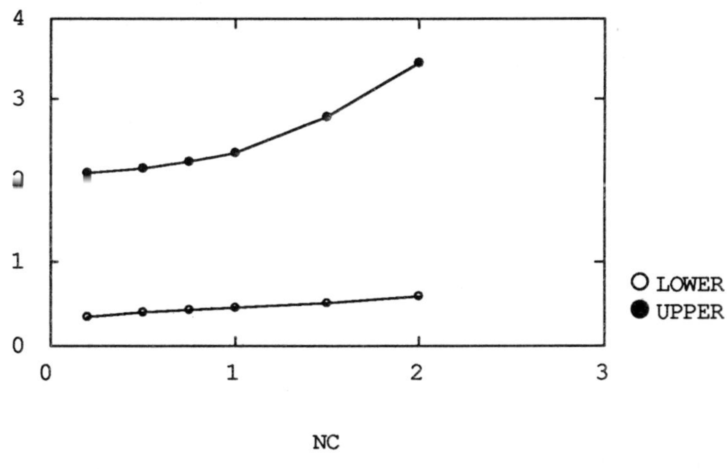

Fig 6.9. Random numbers were generated so that $Y \sim N(-cx_i^2;1)$ for 21 equidistant x-observations in the x-interval $-10 \le x \le 10$. The position of the nearest x-observation compared to x_{max} was randomized. The 95% confidence limits of the distribution of $\hat{\sigma}^2$ were estimated for different values of NC.

It is also obvious that the 95% confidence interval becomes bigger when NC increases. The simulation results are also illustrated in the figure 6.9.

Since the χ^2-distribution in the corresponding situation has 20 degrees of freedom we can compare the obtained results, rescaled as $(N-1)\,\hat{\sigma}^2/\sigma^2$, with the 95% confidence limits of the χ^2-distribution, which are 9.59 and 34.17. The result of this comparison is shown in table 6.14.

TABLE 6.14.

NC	0.20	0.50	0.75	1.00	1.50	2.00	$\chi^2(20)$
lower limit	7.00	8.00	8.50	9.00	10.1	12.0	9.59
upper limit	41.5	43.0	44.5	47.0	55.3	69.0	34.17

Table 6.14. Random numbers were generated so that $Y \sim N(-cx^2;1)$ for equidistant x-observations in the x-interval $-10 \le x \le 10$. The 0.025 and 0.975 limits of the estimated distribution of $\hat{\sigma}^2$, using the proposed estimation method, were rescaled for $0.20 \le NC \le 2.00$ and compared to the χ^2-distribution with 20 df.

The result in this table shows that the lower level corresponds rather well to the χ^2-distribution but the upper limit is much higher in the estimated distributions. It is also obvious that the obtained limits are some functions of NC.

To get an expression for the connection between NC and the estimated variance we use some results of Mammen (1988). He gives an expression, containing $\mu(x)$ and $\hat{\mu}(x)$, that converges to a limit distribution for concave regression. This evaluation is further discussed in chapter 7.

Let

$\mu''(x_0)$ = the curvature in x_0,

$\sigma(x_0)^2 = \mathrm{Var}(Y|X=x_0)$,

$N\,F'(x_0)$ = the density of observations in x_0,

G = the universal limit distribution.

Using these notations we have according to Mammen

$$N^{2/5} \frac{F'(x_0)^{2/5}}{\sigma(x_0)^{4/5}\,\mu''(x_0)^{1/5}} \left(\hat{\mu}(x_0) - \mu(x_0)\right) \rightarrow G.$$

From this expression we see that the distribution depends on the curvature of $\mu(x)$ and the variance of the observations. The product $N^{2/5}\,F'(x_0)^{2/5}$ is used as a scaling factor. The limit distribution G has a very complecated form and Mammen gives no theoretical expression for it. Probably the mathematical expression for how the curvature affects the estimation of the variance is not easy to find.

6.5.3. Properties of the proposed variance estimation method for regression function with piecewise linear functions.

A simulation study was performed for a regression function with piecewise linear functions where

$$\mu(x) = \begin{cases} cx & \text{for } x \leq 0 \\ -cx & \text{for } x > 0. \end{cases}$$

Suppose that we have observations $Y_i = \mu(x_i) + \varepsilon_i$ where $\varepsilon_i \sim N(0;1)$ and the x-observations are obtained equidistantly in the x-interval $-30 \le x \le 30$ with the distance $\Delta = 1$ between two successive x-observations. The position of the nearest x-value compared to x_{max} is uniformly distributed in the interval $(-0.5\Delta, 0.5\Delta)$ centered at x_{max}. In the simulations 25 000 replications were obtained from the above situation and used for the estimation of $Var(\hat{Y}) = \hat{\sigma}^2$ according to the proposed estimation method

$$\hat{\sigma}^2 = \frac{\frac{2}{3}}{N-2} \sum_{i=2}^{N-1} \left[Y_i - \frac{1}{2}(Y_{i-1} + Y_{i+1})\right]^2 - \frac{2}{3}\frac{1}{4}\left(\frac{\hat{\beta}_1 - \hat{\beta}_M}{N-1}\right)^2.$$

The result is presented in table 6.15. The estimated correction term is denoted by corr.:

TABLE 6.15.

$N\beta = \frac{c\Delta}{\sigma/\sqrt{n}}$	$\hat{E}(s^2)$	corr	$\hat{\sigma}^2$	$\hat{\sigma}$
0.05	1.000	0.000	1.000	1.000
0.20	1.000	0.000	1.000	1.000
0.50	1.001	0.000	1.001	1.000
0.75	1.004	0.001	1.003	1.001
1.00	1.008	0.001	1.007	1.003
1.50	1.020	0.001	1.019	1.009
2.00	1.035	0.001	1.036	1.018

Table 6.15. Random numbers were generated so that $Y \sim N(\mu(x);1)$, where $\mu(x)=cx$ for $x \le 0$ and $\mu(x)=-cx$ for $x>0$, for equidistant x-observations in the x-interval $-10 \le x \le 10$. The variances, $\hat{\sigma}^2=\hat{E}(s^2)$ - corr, were estimated according to the proposed estimation method for $0.05 \le NC \le 2.00$.

In the same way as in the preceding study using regression functions with constant curvature we can see that both $\hat{E}(s^2)$, the estimated correction term and $\hat{\sigma}^2$ increase very slowly when Nβ increases. The simulation results show that $\hat{\sigma}^2$ will converge to σ^2. The simulation results indicate that the estimated $Var(\hat{\sigma}^2)$ also increase very slowly when Nβ increases.

When we compare the results from these simulations with the results from the preceding study, we see that the proposed estimation method gives a variance estimate that is closer to σ^2 for the regression functions consisting of piecewise linear functions than for the regression functions with constant curvature. The estimated $Var(\hat{\sigma}^2)$ is also much smaller for regression functions with piecewise linear functions compared to regression functions with constant curvature. In this study we also determined the estimates of the standardized skewness and the standardized kurtosis for different values of the normalized slope, Nβ, together with the estimated distributions to examine the variation in the results. The results were rescaled and shown in table 6.16.

TABLE 6.16.

$N\beta = \frac{c\Delta}{\sigma/\sqrt{n}}$	$\frac{\mu_3}{(\text{stand dev}(\sigma)^3)}$	$\frac{\mu_4}{(\text{stand dev}(\sigma)^4)}$
0.20	0.638	3.592
0.50	0.638	3.593
0.75	0.638	3.597
1.00	0.637	3.595
1.50	0.636	3.595
2.00	0.634	3.591

Table 6.16. Random numbers were generated so that $Y \sim N(\mu(x);1)$, where $\mu(x)=cx$ for $x \leq 0$ and $\mu(x)=-cx$ for $x>0$ for equidistant x-observations in the interval $-10 \leq x \leq 10$. The standardized skewness and the standardized kurtosis were obtained for the variance estimates for $0.20 \leq NC \leq 2.00$.

From this table we can see that any tendencies of the standardized skewness or the standardized kurtosis are not possible to detect in this study. Compared to the distribution of the estimated variance using regression functions with constant curvature both the standardized skewness and the standardized kurtosis were smaller in the examined situation.

6.5.4. Properties of the proposed variance estimation method for a skew regression function.

A simulation study was performed to examine the proposed variance estimation method when the regression function had a small skewness. We used 25 000 replications for the regression function $Y_i = \mu(x_i) + \varepsilon_i$, where $\mu(x) = -cx^2 + dx^3$ and $\varepsilon_i \sim N(0;1)$ for $-10 \le x \le 3$. The x-observations were chosen equidistantly. The position of the nearest x-value compared to x_{max} was uniformly distributed in the interval (-0.5Δ, 0.5Δ). The variation interval for c was 0.5 - 2.00 and the variation interval for d was 0.000 - 0.100. The results from the simulations were rescaled and shown in table 6.17.

TABLE 6.17.

d = 0.000

c	$\hat{E}(s^2)$	corr	$\hat{\sigma}^2$	$\hat{\sigma}$
0.50	1.1673	0.1238	1.0434	1.021
1.00	1.6672	0.4827	1.1832	1.088
1.50	2.5004	1.1525	1.3463	1.160
2.00	3.6587	2.1626	1.5027	1.226

$d = 0.025$

C	$\hat{E}(s^2)$	corr	$\hat{\sigma}^2$	$\hat{\sigma}$
0.50	1.3960	0.2455	1.1498	1.072
1.00	2.0458	0.7218	1.3226	1.150
1.50	3.0251	1.5289	1.4984	1.224
2.00	4.3340	2.6815	1.6616	1.289

$d = 0.050$

C	$\hat{E}(s^2)$	corr	$\hat{\sigma}^2$	$\hat{\sigma}$
0.50	1.7821	0.3987	1.3823	1.176
1.00	2.5819	0.9915	1.5889	1.260
1.50	3.7064	1.9212	1.7921	1.339
2.00	5.1671	3.2045	1.9698	1.404

$d = 0.100$

C	$\hat{E}(s^2)$	corr	$\hat{\sigma}^2$	$\hat{\sigma}$
0.50	3.0229	0.7535	2.2718	1.507
1.00	4.1158	1.5803	2.5444	1.595
1.50	5.5440	2.8182	2.7362	1.654
2.00	7.3071	4.4087	2.9004	1.703

Table 6.17. Random numbers were generated so that $Y \sim N(\mu(x);1)$, where $\mu(x)=-cx^2+dx^3$ for equidistant x-observations in the x-interval $-10 \le x \le 3$. The variances, $\hat{\sigma}^2=\hat{E}(s^2)$ - corr, were estimated according to the proposed estimation method for $0.50 \le C \le 2.00$.

The estimated correction term is denoted by corr. The choice of the variation intervals for c and d was dependent of the fact that for d>0 $\mu(x)$ is a third-degree function, which has a minimum in x = 2c/3d. For bigger x-values we have no longer a unimodal function. In this situation the chosen values of c and d give us this minimum outside the examined x-interval [-10,3].
This study shows that $\hat{E}(s^2)$, the correction term and $\hat{\sigma}^2$ increase as the value of d increases. We can also see that even for such a small skewness as d=0.05, together with a moderate value of c, the variance estimation method gives us estimates that overestimate the variance rather much. In this study we also determined the estimates of the standardized skewness and the standardized kurtosis for different values of c in connection with different values of d to examine the variation in the results. This is shown in table 6.18 and 6.19.

TABLE 6.18. $\dfrac{\mu_3}{\text{stand dev}(\sigma^3)}$

c \ d	0.000	0.025	0.050	0.100
0.50	1.222	1.166	0.998	0.490
1.00	1.100	1.017	0.878	0.691
1.50	1.126	1.171	1.200	1.225
2.00	1.581	1.805	1.944	2.026

Table 6.18. Random numbers were generated so that Y ~ N($\mu(x)$;1), where $\mu(x)=-cx^2+dx^3$ for equidistant x-observations in the x-interval $-10 \leq x \leq 3$. The standardized skewness was obtained for the variance estimates for $0.50 \leq c \leq 2.00$ and $0.00 \leq d \leq 0.10$.

The simulation results indicate that when d increases the standardized skewness decreases. When c increases from 0.00 to 1.00 the standardized skewness decreases. For c>1 skewness increases again. We can also see that when c=0.50 and c=1.00 and d increases, $\mu_3/(\text{stand dev}(\sigma^3))$ decreases. When c=1.50 and c=2.00 and d increases, $\mu_3/(\text{stand dev}(\sigma^3))$ increases again.

TABLE 6.19. $\dfrac{\mu_4}{\text{stand dev}(\sigma^4)}$

c \ d	0.000	0.025	0.050	0.100
0.50	5.190	5.042	4.844	3.856
1.00	4.727	4.402	3.809	2.885
1.50	4.394	4.328	4.077	3.670
2.00	5.887	6.665	6.751	6.572

Table 6.19. Random numbers were generated so that $Y \sim N(\mu(x);1)$, where $\mu(x)=-cx^2+dx^3$ for equidistant x-observations in the x-interval $-10 \leq x \leq 3$. The standardized kurtosis was obtained for the variance estimates for $0.50 \leq c \leq 2.00$ and $0.00 \leq d \leq 0.10$

In this table we can see tendencies that look very much like the tendencies of the skewness. When $c \leq 1.50$ and d increses the standardized kurtosis decreases. When d is fixed and c increases from 0.50 to 1.50 the kurtosis decreases. When $c > 1.50$ the kurtosis increases again.

6.5.5. Properties of the proposed variance estimation method for sigmoid regression functions.

A simulation study was performed to examine the proposed variance estimation method for sigmoid regression functions. In the study 25 000 replications were used for the regression function $Y_i = \mu(x_i) + \varepsilon_i$, where

$$\mu_1(x) = a+10 * \frac{e^x}{1 + e^x}$$

$$\mu_2(x) = a +10*(1-e^{-e^x})$$

and $\varepsilon_i \sim N(0;1)$ for $-10 \le x \le 10$. The x-observations were chosen equidistantly. The distance between two successive x-observations, Δ, varied between 0.67 and 2.00. The position of the nearest x-value compared to x_{inf} was uniformly distributed in the x-interval $(-0.5\Delta, 0.5\Delta)$. The results from the simulations were rescaled and shown in table 6.20. The estimated correction term will in this situation have a different form, namely

$$\frac{2}{3}\,\frac{1}{4}\,\left[\frac{(\hat\beta_{inf} - \hat\beta_1) + (\hat\beta_{inf} - \hat\beta_M)}{N}\right]^2 = \frac{2}{3}\,\frac{1}{4}\,\left[\frac{2\hat\beta_{inf} - (\hat\beta_1 + \hat\beta_M)}{N}\right]^2.$$

where $\hat\beta_{inf}$ is the notation of the slope of the estimated inflection interval.

TABLE 6.20.

N	Δ	$\mu_1(x)$ $\hat{E}(s^2)$	corr	$\hat\sigma^2$	$\hat\sigma$	$\mu_2(x)$ $\hat{E}(s^2)$	corr	$\hat\sigma^2$	$\hat\sigma$
11	2.00	1.326	0.036	1.290	1.136	1.757	0.053	1.704	1.305
16	1.33	1.077	0.024	1.053	1.026	1.219	0.036	1.183	1.088
21	1.00	1.025	0.018	1.008	1.004	1.081	0.027	1.054	1.027
26	0.80	1.011	0.014	0.997	0.999	1.037	0.021	1.016	1.008
31	0.67	1.004	0.012	0.992	0.996	1.017	0.017	1.000	1.000

Table 6.20. Random numbers were generated so that $Y \sim N(\mu_i(x);1)$, where $\mu_1(x)$ is the logit function and $\mu_2(x)$ is the extreme-value function, for equidistant x-observations in the x-interval $-10 \le x \le 10$. The variances, $\hat\sigma^2 = \hat{E}(s^2)$ - corr, were estimated according to the proposed estimation method for $11 \le N \le 31$.

From this table we can see that the estimated variances are smaller for the symmetric logit function than for the non-symmetric extreme-value function. It is also obvious that for the two examined regression functions both $\hat{E}(s^2)$, the correction term, $\hat{\sigma}^2$ and $\hat{\sigma}$ decrease when N increses. For the examined N-values the estimated standard deviations were all very close to σ = 1.00.

This simulation study shows that we have the same tendency for regression functions with constant curvature and regression functions consisting of piecewise linear functions as for sigmoid regression functions. The estimated variance decreases when the number of x-observations in a fixed interval increases or when the distance between two successive x-observations decreases.

TABLE 6.21.

		$\mu_1(x)$		$\mu_2(x)$	
N	Δ	$\frac{\mu_3}{(\text{stand dev}(\sigma)^3)}$	$\frac{\mu_4}{(\text{stand dev}(\sigma)^4)}$	$\frac{\mu_3}{(\text{standdev}(\sigma)^3)}$	$\frac{\mu_4}{(\text{stand dev}(\sigma)^4)}$
11	2.00	1.290	5.469	1.232	5.335
16	1.33	1.106	4.759	1.063	4.633
21	1.00	1.008	4.496	0.982	4.405
26	0.80	0.886	4.166	0.884	4.166
31	0.67	0.824	4.021	0.823	4.019

Table 6.21. Random numbers were generated so that Y ~ $N(\mu_i(x);1)$, where $\mu_1(x)$ is the logit function and $\mu_2(x)$ is the extreme-value function, for equidistant x-observations in the x-interval $-10 \leq x \leq 10$. The standardized skewness and the standardized kurtosis were obtained for the variance estimates for $11 \leq N \leq 31$.

In this study we also determined the estimates of $\mu_3/(\text{stand.dev}(\sigma)^3)$ and $\mu_4/(\text{stand.dev}(\sigma)^4)$ for different values of N together with the estimated distributions to examine the variation in the results. The results were rescaled and shown in table 6.21.

From this table we can see that for the two examined regression functions both the standardized skewness the standardized kurtosis decrease when N increases. If we compare the logit function to the extreme-value function we can see that both the standardized skewness and the standardized kurtosis seem to be smaller in the situation using the non-symmetric regression function.

7. THE DISTRIBUTION OF SOME ESTIMATORS.

As a complement to bias and variance properties it is often interesting to get some idea of the shape of the distribution of the examined estimator. In this chapter we will examine how the distributions of $\hat{Y}(X=x_i)$ are affected by the estimation method depending on $\hat{Y}_i$ being in the middle of the x-interval or near one of the end points. This is done by a simulation study where the 3:rd, μ_3, and the 4:th, μ_4, moments are determined for different positions of $\hat{Y}_i$. For the situation with concave regression we are also interested to see if the distribution of $\hat{Y}(X=x_i)$ is affected by change of curvature for the regression function with constant curvature. This is also examined by simulations. For regression functions with constant curvature we also try to estimate a limit distribution of $\hat{Y}(X=x_i)$ using theoretical results concerning concave/convex regression obtained by Mammen (1988). In this situation we use the Egdeworth expansion. For the distributions of $\hat{Y}_{max}$ and $\hat{X}_{max}$ in concave-unimodal regression and the distributions of $\hat{Y}_{inf}$ and $\hat{X}_{inf}$ in sigmoid regression these calculations would be too complicated.

For these four estimators we have examined the distributions approximately by determining the 3:rd and 4:th moments for different situations in simulations.

7.1. The distribution of $\hat{Y}|X$

In some situations we are not interested in the individual values of $\mu(x)$ but rather on the shape of the regression function. In the simulation studies described in the chapters 5.2 and 6.2 we could see that bias and the variance of the estimates were dependent on there being a regression function with constant curvature or a regression function using linear functions. As a complement to these studies we also examine how the distribution of $\hat{Y}(X=x_i)$ is affected by these two situations.

7.1.1. The distribution of $\hat{Y}|X$ for increasing and concave regression functions with constant curvature.

To examine the shape of the distribution of $\hat{Y}(X=x_i)$ for regression functions with constant curvature a simulation study was performed according to the following:

We generate a sample $y_i = \mu(x_i) + \varepsilon_i$, i=1,....,N, where $\mu(x_i)$ is an isotonic-concave regression function with equidistant x-observations in the interval $-d \leq x \leq 0$ and $\varepsilon_i \sim N(0; \sigma(x_i)^2)$, for some $\sigma(x_i)^2$, $0< \sigma(x_i)^2 < \infty$, i=1,....,N. The examined regression functions was $\mu(x)=-x^2$. The simulation study was performed for

1) 16 equidistant x-observations in the interval $-5 \leq x \leq 0$, ie the distance between two successive x-observations was $\Delta=1/3$.
2) $m(x_i)=1$, i=1,...,16.
3) $\varepsilon_i \sim N(0; 1)$.

The standardized 3:rd, μ_3/σ^3, and standardized 4:th moments, μ_4/σ^4, were estimated for different y-values. The results from 25 000 replications are shown in table 7.1.

TABLE 7.1. x_i	$\frac{\mu_3}{s^3}$	$\frac{\mu_4}{s^4}$
-5.00	-0.177	4.238
-4.67	0.228	6.817
-4.33	0.283	3.140
-4.00	0.307	5.381
-3.67	0.300	3.646
-3.33	0.322	3.233
-3.00	0.293	3.142
-2.67	0.285	3.084
-2.33	0.318	3.145
-2.00	0.336	3.223
-1.67	0.311	3.174
-1.33	0.282	3.060
-1.00	0.202	3.001
-0.67	0.140	3.085
-0.33	0.081	3.009
0.00	0.225	3.026

Table 7.1. Random numbers were generated so that $Y \sim N(-x_i^2;1)$ for 16 equidistant x-observations in the x-interval $-5 \le x \le 0$. The standardized 3:rd and the standardized 4:th moments of $\hat{Y}_i$ were estimated for different x-values.

From this simulation study we can see that the values of the standardized skewness, μ_3/σ^3, were not especially close to zero, ie the distributions were skew, and the standardized kurtosis, μ_4/σ^4, was bigger near the end points than in the middle. But all estimates were bigger than the corresponding values in the normal distribution. Thus even if we adjust for bias we can´t use the normal distribution when we construct confidence interval. In order to get a better idea of the shape of the distributions we use the results in Mammen (1988). Mammen theoretically studies an expression, containing $\mu(x)$ and $\hat{\mu}(x)$, which converges to a limit distribution for concave regression according to the following:

Let $Y_i = \mu(x_i) + \varepsilon_i$, i=1,2,.....,N, be observations at points x_i, i=1,2,.....,N. The ε_i:s are supposed to be independent random variables with $E(\varepsilon_i) = 0$ and bounded variances depending on x_i only. It is also supposed that the observation points are chosen in such a way that their distribution converges to F(x), which is differentiable.

Let

$\mu''(x_0)$ = the curvature in x_0,

$\sigma(x_0)^2 = Var(Y|X=x_0)$,

$N\,F'(x_0)$ = the density of observations in x_0.

Observe that $F(x_0)$ and $\mu'(x_0)$ depend on N. Using these notations Mammen shows that

$$C^* = N^{2/5} \frac{F'(x_0)^{2/5}}{\sigma(x_0)^{4/5}\,\mu''(x_0)^{1/5}} \left(\hat{\mu}(x_0) - \mu(x_0) \right) \rightarrow G,$$

where G is a universal limit distribution defined in a certain way for the Brownian motion process.

From this expression we see that the distribution of $\hat{Y}|X$ depends on the curvature of $\mu(x)$ and the variance of the Y-observations, $Var(Y|X=x_0)$. The product $N^{2/5}\,F'(x_0)^{2/5}$ acts as a scaling factor. The limit distribution G has a very complicated form and Mammen gives no closed expression for it. To get a picture of the shape of the distribution G simulations were made using the same assumptions as in the preceding simulation study in this chapter. In these assumptions we had the same curvature on the whole curve and

$Var(Y|X=x_0)$ was the same for each x_0. Thus the limit distribution, G, is the same for every x-value.

The estimates of Y|X using 25 000 replications were grouped in 50 equally sized groups and $F(x_0)$ was estimated. Since the estimation method has some boundary affects $\hat{\mu}(x_0)$ has the best estimate for one of the central x-values. For this purpose x_0=-2.667 is chosen. From the simulations we obtain $\hat{E}(\hat{Y}| x_0=-2.667) = -7.119$, while $\mu(x=-2.667) = -7.111$, ie a very small bias. These simulation results indicate that the distribution of $\hat{\mu}(x_0)$ has only little skewness. This is confirmed by the estimated standardized skewness, μ_3/s^3, which was 0.285 according to the simulations. The corresponding standardized kurtosis, μ_4/s^4, was estimated to 3.084. The variance was estimated to $\hat{Var}[\hat{\mu}(x=-2.667)] = 0.270$. For the point x=-2.667 the normalizing constant

$$C^* = N^{2/5} \frac{F'(x_0)^{2/5}}{\sigma(x_0)^{4/5}\ \mu''(x_0)^{1/5}}$$

is obtained by:

$N^{2/5}\,F'(x_0)^{2/5} = (\frac{15}{5})^{2/5} = 1.55184.$

$\sigma(x_0)^{4/5} = 1$ according to the assumptions and

$\mu''(x_0)^{1/5} = (2)^{1/5} = 1.148698$ according to the assumptions.

Thus the normalizing constant, C^*, equals 1.351. Denote the estimated cumulative distribution by G_{16} since we have 16 x-observations in the examined x-interval. This means that G_{16} is estimated by the empirical cumulative distribution of $v(x=-2.667)$ = $1.351 * [\hat{\mu}(x=-2.667)-\mu(x=-2.667)]$ in the simulations. The corresponding frequency function is denoted by g_{16}. The results obtained from the simulation study is shown in figure 7.1.

FIGURE 7.1.

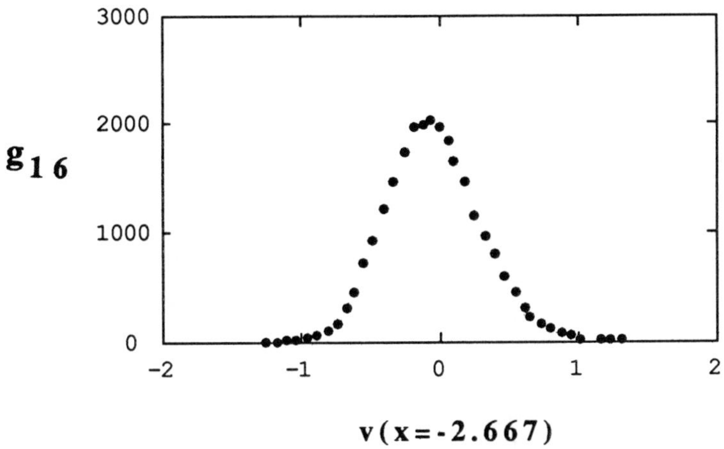

Fig 7.1. Random numbers were generated so that Y ~ N(- x_i^2;1) for 16 equidistant x-observations in the x-interval -5 $\le x \le 0$. The distribution of $\hat{\mu}(x)$ for x=-2.667 was determined. The universal limit distribution G under these assumptions was estimated by $\nu(x)$ = 1.351 * [$\hat{\mu}$(x=-2.667) - μ(x=-2.667)]. The corresponding frequency function g_{16} was determined.

The simulations were repeated for the situation that the number of x-observations were increased from 16 to 26 in the x-interval -5 $\le$ x $\le$ 0. Thus the scaling factor is changed from $(15/5)^{2/5}$ = 1.55184 to $(25/5)^{2/5}$ = 1.90365. The normalizing constant, C^*, will in this situation be 1.657. The limit distribution was estimated by the distribution obtained for x_0 = -2.600.

Denote the estimated cumulative distribution by G_{26} since we have 26 x-observations in the examined x-interval. This menas that G_{26} is estimated by the empirical distribution of ν(x=-2.600)=1.657*

[$\hat{\mu}$(x=-2.600)- μ(x=-2.600)] in the simulations. The results obtained from the two simulation studies are shown in table 7.2.

TABLE 7.2.

ν(x)	G_{16}	G_{26}	ν(x)	G_{16}	G_{26}
-2.3	0.0001	0.0000	0.5	0.7742	0.7740
-2.2	0.0003	0.0001	0.6	0.8134	0.8127
-2.1	0.0007	0.0003	0.7	0.8456	0.8464
-2.0	0.0009	0.0006	0.8	0.8734	0.8753
-1.9	0.0010	0.0011	0.9	0.8976	0.9012
-1.8	0.0021	0.0019	1.0	0.9190	0.9205
-1.7	0.0038	0.0036	1.1	0.9362	0.9369
-1.6	0.0062	0.0063	1.2	0.9514	0.9501
-1.5	0.0096	0.0101	1.3	0.9626	0.9619
-1.4	0.0155	0.0153	1.4	0.9707	0.9708
-1.3	0.0236	0.0229	1.5	0.9776	0.9784
-1.2	0.0350	0.0335	1.6	0.9827	0.9838
-1.1	0.0509	0.0487	1.7	0.9872	0.9877
-1.0	0.0717	0.0678	1.8	0.9908	0.9912
-0.9	0.0960	0.0903	1.9	0.9935	0.9936
-0.8	0.1264	0.1208	2.0	0.9952	0.9953
-0.7	0.1622	0.1566	2.1	0.9965	0.9966
-0.6	0.2032	0.1968	2.2	0.9976	0.9977
-0.5	0.2487	0.2433	2.3	0.9982	0.9984
-0.4	0.3000	0.2946	2.4	0.9988	0.9987
-0.3	0.3548	0.3481	2.5	0.9993	0.9991
-0.2	0.4116	0.4062	2.6	0.9995	0.9994
-0.1	0.4694	0.4653	2.7	0.9996	0.9996
0.0	0.5256	0.5207	2.8	0.9998	0.9998
0.1	0.5810	0.5786	2.9	0.9998	0.9999
0.2	0.6349	0.6326	3.0	0.9999	0.9999
0.3	0.6846	0.6827	3.1	0.9999	1.0000
0.4	0.7309	0.7319	3.2	1.0000	1.0000

Table 7.2. Random numbers were generated so that Y ~ N($-x_i^2$;1) for 16 respectively 26 equidistant x-observations in the x-interval $-5 \leq x \leq 0$. The distributions of $\hat{\mu}$(x=-2.667) using 16 x-observations and of $\hat{\mu}$(x=-2.600) using 26 x-observations were estimated. The universal limit distribution G under these assumptions was estimated for these two situations.

We also make an estimated Edgeworth expansion for the limiting distribution. Since the Edgeworth expansion use the standardized variable we also transform it according to:
Let U = $\hat{\mu}(x=-2.667)$ - $\mu(x=-2.667)$. We can standardize U to Z = $(U-\mu(U))/\sqrt{Var(U)}$. The estimates of $\mu(U)$ and $\sqrt{Var(U)}$ are obtained by the simulations where $\mu(U)$ = -0.008 and $\sqrt{Var(U)}$ = 0.270. From these simulation results we can give an approximation of Z according to:

$$Z = \frac{\hat{\mu}(x=-2.667)-\mu(x=-2.667)-0.008}{\sqrt{\hat{V}ar(x=-2.667)}} = \frac{\hat{\mu}(x=-2.667)-7.119}{0.270}$$

The obtained $\nu(x)$ is also shown in table 7.3 together with the estimated cumulative distribution, $\hat{G}$, obtained by the simulation study. In table 7.3 we also give the corresponding smoothed distribution, $\hat{G}_{Edg}$, obtained by using the Edgeworth expansion, see e.g. Bickel & Doksum (1977), according to the following:

Let

$$H_k(x) = \frac{(-1)^k \, d^k\phi(x)}{\phi(x) \, d^k x} \qquad k=0, 1, 2.............$$

be the Hermite polynomials of degree k, where $\phi(x)$ is the density function of a standard normal variable. The first six of these polynomials are

$$H_0(x)=1, \quad H_1(x)= x, \quad H_2(x)= x^2 - 1, \quad H_3(x)= x^3 - 3x,$$
$$H_4(x)= x^4 - 6x^2 + 3, \quad H_5(x)= x^5 - 10x^3 + 15x.$$

Using these orthogonal polynomials we can construct the Edgeworth expansion

TABLE 7.3.

Z(x)	v(x)	$\hat{G}$	$\hat{G}_{Edg}$	F_{Normal}	Z(x)	v(x)	$\hat{G}$	$\hat{G}_{Edg}$	F_{Normal}
-3.4	-1.240	0.0000	0.0000	0.0003	0.6	0.219	0.7429	0.7351	0.7257
-3.3	-1.204	0.0000	0.0000	0.0005	0.7	0.255	0.7726	0.7649	0.7580
-3.2	-1.167	0.0000	0.0000	0.0007	0.8	0.292	0.8013	0.7926	0.7881
-3.1	-1.131	0.0001	0.0000	0.0010	0.9	0.328	0.8254	0.8181	0.8159
-3.0	-1.094	0.0002	0.0000	0.0013	1.0	0.365	0.8476	0.8414	0.8413
-2.9	-1.058	0.0003	0.0002	0.0019	1.1	0.401	0.8685	0.8624	0.8643
-2.8	-1.021	0.0004	0.0004	0.0026	1.2	0.438	0.8862	0.8814	0.8849
-2.7	-0.985	0.0010	0.0007	0.0035	1.3	0.474	0.9020	0.8984	0.9032
-2.6	-0.948	0.0016	0.0012	0.0047	1.4	0.511	0.9178	0.9133	0.9192
-2.5	-0.912	0.0026	0.0019	0.0062	1.5	0.547	0.9305	0.9266	0.9332
-2.4	-0.875	0.0038	0.0031	0.0082	1.6	0.584	0.9416	0.9381	0.9452
-2.3	-0.839	0.0055	0.0047	0.0107	1.7	0.620	0.9520	0.9480	0.9554
-2.2	-0.803	0.0079	0.0071	0.0139	1.8	0.657	0.9596	0.9567	0.9641
-2.1	-0.766	0.0100	0.0102	0.0179	1.9	0.693	0.9666	0.9640	0.9713
-2.0	-0.730	0.0143	0.0144	0.0228	2.0	0.730	0.9719	0.9702	0.9772
-1.9	-0.693	0.0191	0.0195	0.0298	2.1	0.766	0.9766	0.9755	0.9821
-1.8	-0.657	0.0265	0.0266	0.0359	2.2	0.803	0.9806	0.9800	0.9861
-1.7	-0.620	0.0344	0.0351	0.0446	2.3	0.839	0.9841	0.9837	0.9893
-1.6	-0.584	0.0448	0.0455	0.0558	2.4	0.875	0.9872	0.9868	0.9918
-1.5	-0.547	0.0590	0.0581	0.0668	2.5	0.912	0.9898	0.9893	0.9938
-1.4	-0.511	0.0738	0.0730	0.0808	2.6	0.948	0.9919	0.9914	0.9953
-1.3	-0.474	0.0914	0.0904	0.0968	2.7	0.985	0.9937	0.9931	0.9965
-1.2	-0.439	0.1109	0.1105	0.1151	2.8	1.021	0.9950	0.9945	0.9974
-1.1	-0.401	0.1336	0.1332	0.1357	2.9	1.058	0.9959	0.9956	0.9981
-1.0	-0.365	0.1592	0.1586	0.1587	3.0	1.094	0.9972	0.9966	0.9987
-0.9	-0.328	0.1868	0.1867	0.1841	3.1	1.131	0.9976	0.9973	0.9990
-0.8	-0.292	0.2180	0.2173	0.2119	3.2	1.167	0.9980	0.9979	0.9993
-0.7	-0.255	0.2502	0.2502	0.2420	3.3	1.204	0.9984	0.9984	0.9995
-0.6	-0.219	0.2872	0.2852	0.2743	3.4	1.240	0.9988	0.9988	0.9997
-0.5	-0.182	0.3243	0.3218	0.3085	3.5	1.277	0.9990	0.9991	0.9998
-0.4	-0.146	0.3655	0.3601	0.3446	3.6	1.313	0.9994	0.9992	0.9998
-0.3	-0.109	0.4048	0.3992	0.3821	3.7	1.350	0.9996	0.9995	0.9999
-0.2	-0.073	0.4446	0.4390	0.4207	3.8	1.386	0.9996	0.9996	0.9999
-0.1	-0.036	0.4852	0.4791	0.4602	3.9	1.423	0.9998	0.9998	1.0000
0.0	0.000	0.5256	0.5190	0.5000	4.0	1.459	0.9998	0.9998	1.0000
0.1	0.036	0.5653	0.5582	0.5398	4.1	1.496	0.9998	0.9999	1.0000
0.2	0.073	0.6038	0.5967	0.5793	4.2	1.532	0.9999	0.9999	1.0000
0.3	0.109	0.6409	0.6337	0.6179	4.3	1.569	0.9999	0.9999	1.0000
0.4	0.146	0.6773	0.6693	0.6554	4.4	1.605	0.9999	0.9999	1.0000
0.5	0.182	0.7090	0.7032	0.6915	4.5	1.641	1.0000	1.0000	1.0000

Table 7.3. Random numbers were generated so that $Y \sim N(-x_i^2;1)$ for 16 equidistant x-observations in the x-interval $-5 \leq x \leq 0$. Denote $\hat{\mu}(x=-2.667) - \mu(x=-2.667)$ by U. Th corresponding values of $v(x) = 1.351*U$ were calculated. U was standardized to Z. The universal limit distribution G was estimated. The corresponding Edgeworth expansion and the normal approximation were determined.

$$\hat{F}_{Edg}\hat{G}(x) = \Phi(x) - \phi(x)\left[\frac{1}{6}\left(\frac{\mu_3}{\sigma^3}\right)H_2(x) + \frac{1}{24}\left(\frac{\mu_4}{\sigma^4} - 3\right)H_3(x) + \frac{1}{72}\left(\frac{\mu_3}{\sigma^3}\right)^2 H_5(x)\right] + r_n \qquad \text{where } r_n = o(n^{-1/2}).$$

The normal approximation, F_{Normal}, of the distribution is also illustrated in table 7.3.

7.1.2. The distribution of $\hat{Y}|X$ for increasing and concave regression functions for linear functions.

To be able to compare the simulation results in the study in this chapter to the results in the preceding chapter 7.1.1 containing situations with constant curvature, the same situation as for bias and the variance of the estimates described in the chapters 5.2.2 and 6.2.2 were examined.

Suppose we have a sample $y_k = \mu(x_k) + \varepsilon_k$, k=1,....,N, where $\mu(x_k)$ is an isotonic-concave regression function with equidistant x-observations in the interval $-d \le x \le 0$ and $\varepsilon_k \sim N(0, \sigma(x_k)^2)$, where $\sigma(x_k)^2$ are bounded. The regression functions examined were of the type $\mu(x) = cx$ where the constant c varied from 0.5 to 3.0.

In the simulations with the regression function $\mu(x) = cx$, the normalized slope, $N\beta$, can be expressed in terms of the slope (c), the distance between two successive x-observations (Δ), the variance (σ^2) and the number of y-observations (n) for each x_k by $N\beta = c\Delta\sqrt{n}/\sigma$. The simulation study was performed for

1) 21 equidistant x-observations in the interval $-20 \le x \le 0$, ie the distance between two successive x-observations was $\Delta=1.0$.

2) $n(x_k)=1,\ k=1,...,21.$

3) $\varepsilon_k \sim N(0, 1).$

In table 7.4 the result from 25 000 replications shows the normalized skewness, μ_3/s^3, and the normalized kurtosis, μ_4/s^4, where s is the notation for the standard deviation of the estimate.

TABLE 7.4. $\frac{\mu_3}{s^3}$ and $\frac{\mu_4}{s^4}$

	$\frac{\mu_3}{s^3}$, $N\beta = \frac{c\Delta}{\sigma/\sqrt{n}}$			$\frac{\mu_4}{s^4}$, $N\beta = \frac{c\Delta}{\sigma/\sqrt{n}}$		
x_i	0.25	0.50	0.75	0.25	0.50	0.75
-20.0	-0.602	-0.604	-0.605	3.581	3.590	3.593
-19.0	-0.134	-0.132	-0.131	3.111	3.113	3.114
-18.0	0.005	0.006	0.007	3.086	3.086	3.085
-17.0	0.074	0.074	0.074	3.095	3.091	3.091
-16.0	0.107	0.106	0.105	3.108	3.103	3.102
-15.0	0.148	0.146	0.145	3.162	3.157	3.155
-14.0	0.190	0.187	0.186	3.209	3.206	3.203
-13.0	0.230	0.225	0.224	3.254	3.251	3.250
-12.0	0.257	0.250	0.249	3.271	3.269	3.269
-11.0	0.280	0.272	0.270	3.307	3.303	3.303
-10.0	0.302	0.297	0.294	3.392	3.399	3.397
-9.0	0.306	0.309	0.304	3.444	3.470	3.462
-8.0	0.278	0.288	0.283	3.406	3.431	3.422
-7.0	0.230	0.253	0.247	3.288	3.333	3.319
-6.0	0.169	0.213	0.210	3.169	3.245	3.232
-5.0	0.111	0.159	0.162	3.086	3.145	3.141
-4.0	0.063	0.109	0.119	3.030	3.073	3.084
-3.0	0.027	0.049	0.068	3.023	3.031	3.047
-2.0	-0.001	-0.017	-0.008	3.022	3.058	3.059
-1.0	-0.015	-0.111	-0.113	2.997	3.095	3.134
0.0	0.016	-0.152	-0.249	2.936	2.970	3.009

Table 7.4. Random numbers were generated so that $Y \sim N(\mu(x);1)$ for 21 equidistant x-observations in the x-interval $-20 \le x \le 0$, where $\mu(x)=cx$ for $x<0$ and $\mu(x)=-cx$ for $x\ge0$. The standardized 3:rd and the standardized 4:th moments of $\hat{Y}_i$ were estimated for different values of the normalized slope, $N\beta$.

Since we obtained analogous results of μ_3/s^3 and μ_4/s^4 for all examined values of the normalized slope we only illustrate the result for Nβ = 0.25, 0.50 and 0.75. From table 7.4 we can see that the estimation procedure gives smaller values of μ_3/s^3 at both ends of the interval $-20 \leq x \leq 0$ than in the middle. From the table we can also see that both μ_3/s^3 and μ_4/s^4 are slowly varying in the same way as for bias in the corresponding simulation study .

7.2. The distribution of the maximum point in concave-unimodal regression.

In a unimodal regression function the maximum point is one of the most interesting points. Therefore this section will deal with problems concerning the distribution of the estimated maximum point. Simulation studies have been performed for the situations when the curvature of the regression function is constant and for regression functions with piecewise linear functions. We have also examined how a small skewness affects the distributions of some estimators.

7.2.1. The influence on the distribution of $\hat{Y}_{max}$ by the curvature of the regression function.

To examine the distributions in different situations we start by performing two simulation studies under the restriction that the regression function is unimodal. The first study deals with regression functions having constant curvature. The other study is performed for regression functions with picewise linear functions. These two studies have been made as similar as possible in order to compare the obtained normalized skewness and the normalized kurtosis in the simulations.

7.2.1.1. The distribution of $\hat{Y}_{max}$ for regression functions with constant curvature.

Suppose we have a sample $y_{ij} = \mu(x_i) + \varepsilon_{ij}$, $j=1,....,n(x_i)$, for each of N fixed equidistant x-values, where $\mu(x_i)$ is a symmetric unimodal regression function with maximum in $x = 0$. Let $\varepsilon_i \sim N(0, \sigma(x_i)^2)$, where $\sigma(x_i)^2$ are bounded.

FIGURE 7.2.

The distribution of $\hat{Y}_{max}$

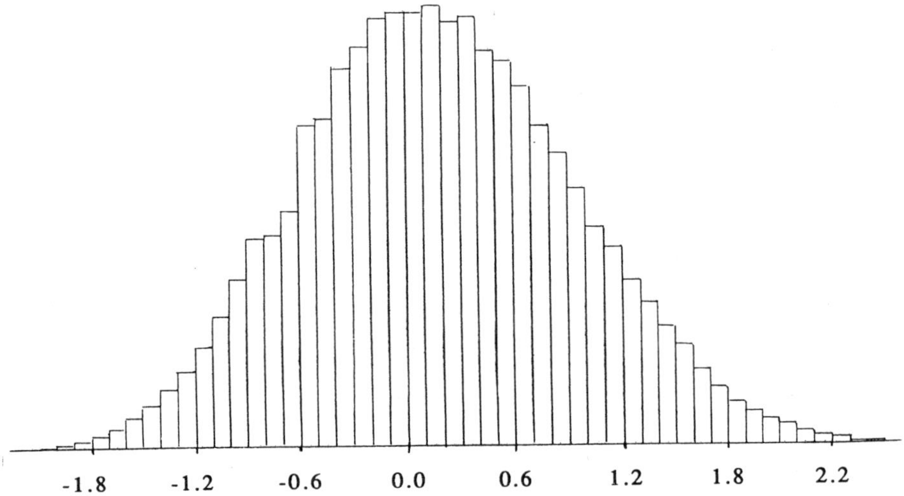

Fig 7.2. Random numbers were generated so that $Y \sim N(-cx_i^2;1)$ for 21 equidistant x-observations in the x-interval $-10 \leq x \leq 10$. The position of the nearest x-value was uniformly distributed in the interval $(-0.5\Delta,0.5\Delta)$ centered at x_{max}. The distribution of $\hat{Y}_{max}$ for c=1 was estimated using 25 000 replications.

To examine how the curvature of the regression function influences the distribution of $Y(\hat{X}_{max})$ a simulation study was performed. The regression functions examined were of the type $\mu(x) = -cx^2$ with constant curvature c. The curvature was chosen to be $0.05 \leq c \leq 3.25$ in different simulations. The corresponding x-interval was $-10 \leq x \leq 10$. The position of the nearest x-value was uniformly distributed in the interval $(-0.5\Delta, 0.5\Delta)$ centered at x_{max}. The distribution for NC=1 is shown in figure 7.2.

From the simulations using 25 000 replications the 3:rd, μ_3, and the 4:th, μ_4, moments for both $\hat{Y}_{max}$ and $\hat{X}_{max}$ were estimated. The standardized skewness μ_3/s^3 and the standardized kurtosis μ_4/s^4 were determined. The results showed that for all values of NC the standardized skewness, μ_3/s^3, for $\hat{Y}_{max}$ was around 0.175 - 0.190 and the standardized kurtosis, μ_4/s^4, very near 3.0. For the distribution shown in figure 7.2 when c=1 the standardized skewness was estimated to μ_3/s^3=0.178 and the standardized kurtosis μ_4/s^4=3.000.

7.2.1.2. The distribution of $\hat{Y}_{max}$ for symmetric regression functions with piecewise linear functions.

This simulation study was performed in the same way as regression functions with constant curvature. In this study we also used N fixed equidistant x-values, where $\mu(x_i)$ was a symmetric unimodal regression function with maximum in x = 0. The regression functions examined were of the type $y_{ij} = \varepsilon_{ij}$, $j=1,\ldots,n(x_i)$, where $\varepsilon_i \sim N(0, \sigma(x_i)^2)$, for bounded $\sigma(x_i)^2$.

In the simulations we used

$$\mu(x) = \begin{cases} cx & \text{for } x \leq 0 \\ -cx & \text{for } x > 0 \end{cases}$$

for the variation interval $0.50 \leq c \leq 2.00$. The corresponding x-interval was $-25 \leq x \leq 25$. The position of the nearest x-value was uniformly distributed in the interval $(-0.5\Delta, 0.5\Delta)$ centered at x_{max}. From the simulations using 25 000 replications the 3:rd, μ_3, and the 4:th, μ_4, moments for $\hat{Y}_{max}$ were estimated. The standardized skewness μ_3/s^3 and the standardized kurtosis μ_4/s^4 were determined. The results showed that for all values of c the standardized skewness, μ_3/s^3, for $\hat{Y}_{max}$ was around -0.314 - -0.227 and the standardized kurtosis, μ_4/s^4, very near 3.0. Compared to the situation with constant curvature we see that in this simulation study the standardized skewness was negative for all values of c while in the case of constant curvature it was positive for all values of NC. When the standardized kurtosis was compared we found that there was no essential difference between the two cases.

7.2.1.3. Comparison between the cumulative distribution of $\hat{Y}_{max}$ and the extreme-value function.

From the preceding simulation studies in chapters 7.2.1.1 and 7.2.1.2 we obtained the estimates, $F_n^{(i,c)}(x)$, i=1,2, of the cumulative distibutions of $\hat{Y}_{max}$ for the regression functions

$$\mu_1(x) = -cx^2 \qquad \text{for } c=0.5,\ 1.0,\ 1.5,\ 2.0,\ 3.0$$

and

$$\mu_2(x) = \begin{cases} cx & \text{if } x < 0.0 \\ -cx & \text{if } x \geq 0.0 \end{cases} \qquad \text{for } c=0.5,\ 1.0,\ 1.5,\ 2.0$$

Since these cumulative distributions are skew we suspect that they could be approximated by extreme-value functions. Thus for each cumulative distribution the values of $F_n^{(i,c)}(x)$, i=1,2, were estimated in twenty equidistant x-observations and for each cumulative distribution an LS estimate of an extreme value function was fitted. The parameters α and β in $\mu(x) = 1 - \exp[-\exp(\alpha x+\beta)]$ were estimated. The estimates are denoted a and b. The results are shown in table 7.5 together with the corresponding value of $\hat{Y}_{max}$ obtained from the simulation study described in chapter 5.3.1.1.

TABLE 7.5.

$\mu_1(x)$				$\mu_2(x)$			
NC	a	b	$\hat{Y}_{max}$	NC	a	b	$\hat{Y}_{max}$
0.5	1.5	-0.5	0.239	0.5	1.5	0.3	-0.467
1.0	1.5	-0.5	0.186	1.0	1.5	0.5	-0.667
1.5	1.4	0.5	0.100	1.5	1.4	0.5	-0.825
2.0	1.0	-0.5	0.059	2.0	1.0	0.5	-0.967
3.0	1.3	-0.4	-0.080				

<u>Table 7.5.</u> Random numbers were generated so that $Y \sim N(-NCx_i^2;1)$ for 21 equidistant x-observations in the x-interval $-10 \le x \le 10$. The cumulative distributions of $\hat{Y}_{max}$ were determined for $0.5 \le NC \le 3.0$. The LS estimates of extreme value functions, $\mu(x) = 1 - \exp[-\exp(ax+b)]$, were fitted to the cumulative distributions for different values of NC.

The differences

$$D_n^{(i,c)} = \sup_{\text{all } x} | F_n^{(i,c)}(x) - \hat{\mu}_i(x) |, \quad i=1,2$$

in the examined situations were calculated and the maximum absolute differences are shown in table 7.6.

If the sample distribution, $F_n^{(i,c)}(x)$, differs from an assumed distribution $\mu_i(x)$ by too much then, according to the Kolmogorov-Smirnov test, this may serve as grounds to reject the hypothesis

TABLE 7.6.

$\mu_1(x)$		$\mu_2(x)$	
NC	$D_n^{(1,c)}$	NC	$D_n^{(2,c)}$
0.5	0.105	0.5	0.172
1.0	0.087	1.0	0.143
1.5	0.060	1.5	0.187
2.0	0.149	2.0	0.205
3.0	0.050		

Table 7.6. Random numbers were generated so that $Y \sim N(\mu_i(x);1)$ for $\mu_1(x)=-cx^2$ using 21 equidistant x-observations in the x-interval $-10 \le x \le 10$ and for $\mu_2(x)=cx$ for $x\le 0$ and $\mu_2(x)=-cx$ for $x>0$ using 51 equidistant x-observations in the x-interval $-25 \le x \le 25$. The cumulative distributions of $\hat{Y}_{max}$ were determined for $0.5 \le NC \le 3.0$. The LS estimates of extreme-value functions were fitted to the cumulative distributions for different values of NC. The maximum absolute differences, $D_n^{(i,c)}$, were determined.

that $\mu_i(x)$ is the population distribution function. If we use the Kolmogorov-Smirnov test we can reject the hypothesis that the distribution of $\hat{Y}_{max}$ comes from a specific extreme value function if $D_n^{(i,c)}$ exceeds 0.231 with a significance level of 0.20. Since this is not the case in any of the examined situations we can conclude that we can not reject the hypothesis that the distribution of $\hat{Y}_{max}$ comes from a specific extreme-value function.

7.2.2. The influence on the distribution of $\hat{X}_{max}$ by the curvature of the regression function.

Another result obtained from the preceding simulation studies was how the distribution of the estimated x_{max} was affected by the curvature of the regression function. In the earlier study of $\hat{Y}_{max}$ we separeted the two cases when the curvature of regression function is constant and when the regression function contains piecewise linear functions. This was also the case with $\hat{X}_{max}$.

7.2.2.1. The distribution of $\hat{X}_{max}$ for regression functions with constant curvature.

In the same simulation study as for $\hat{Y}_{max}$ we also determined the estimates of μ_3/σ^3 and μ_4/σ^4 for $\hat{X}_{max}$. The regression functions examined were of the type $\mu(x) = -cx^2$ for the variation interval $0.05 \leq c \leq 3.25$. The corresponding x-interval was $-10 \leq x \leq 10$. The position of the nearest x-value was uniformly distributed in the interval $(-0.5\Delta, 0.5\Delta)$ centered at x_{max}.

For all values of NC we obtained μ_3/s^3=0.000, which was expected since we have a symmetric situation for the x-values. The results showed that for all values of NC the standardized kurtosis, μ_4/s^4, for $\hat{X}_{max}$ was around 2.500 - 2.660. This means that we have a non-negligible difference compared to the corresponding value of the normal distribution, 3.0.

The distribution for NC=1 is shown in figure 7.3.

FIGURE 7.3.

The distribution of $\hat{X}_{max}$

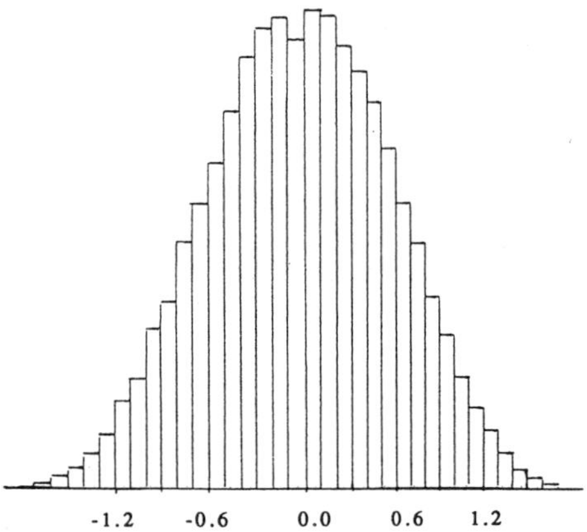

Fig 7.3. Random numbers were generated so that $Y \sim N(-cx_i^2;1)$ for 21 equidistant x-observations in the x-interval $-10 \leq x \leq 10$. The position of the nearest x-value was uniformly distributed in the interval $(-0.5\Delta,\ 0.5\Delta)$ centered at x_{max}. The distribution of $\hat{X}_{max}$ when c=1 was estimated using 25 000 replications.

7.2.2.2. The distribution of $\hat{X}_{max}$ for regression functions with piecewise linear functions.

In the same simulation study as for $\hat{Y}_{max}$ we also determined the estimates of μ_3/σ^3 and μ_4/σ^4 for $\hat{X}_{max}$. In this study we also used N fixed equidistant x-values, where $\mu(x_i)$ was a symmetric unimodal regression function with maximum in x = 0. In the simulations we use

$$\mu(x) = \begin{cases} cx & \text{for } x \le 0 \\ -cx & \text{for } x > 0 \end{cases}$$

for the variation interval $0.50 \le c \le 2.00$. The corresponding x-interval was $-25 \le x \le 25$. The position of the nearest x-value was uniformly distributed in the interval $(-0.5\Delta, 0.5\Delta)$ centered at x_{max}. From the simulations using 25 000 replications the 3:rd, μ_3, and the 4:th, μ_4, moments for $\hat{X}_{max}$ were estimated. The standardized skewness and the standardized kurtosis were determined. Since the simulation situation was symmetric around the true x_{max}, μ_3/σ^3 equals 0.0. This was confirmed by the simulations where the estimate μ_3/s^3 was determined to 0.000. The simulation result for the standardized kurtosis is shown in table 7.7.

TABLE 7.7.

c	0.50	0.75	1.00	1.25	1.50	1.75	2.00
$\left(\frac{\mu_4}{s^4}\right)$	3.970	3.872	3.682	3.481	3.215	3.080	2.859

Table 7.7. Random numbers were generated so that $Y \sim N(\mu(x);1)$ for 51 equidistant x-observations in the x-interval $-25 \le x \le 25$, where $\mu(x)=cx$ for $x \le 0$ and $\mu(x)=-cx$ for $x>0$. The standardized 4:th moments of $\hat{X}_{max}$ were estimated for $0.5 \le c \le 2.0$.

From this table we see that the standardized kurtosis is rather big for the smallest values of c. When c increases the standardized kurtosis decreases. If we compare the results with the corresponding results in the preceding chapter we see that the standardized kurtosis is bigger for regression functions with piecewise linear functions compared to regression functions with constant curvature.

7.2.3. The influence on the distributions of some estimators by the skewness of the regression function.

Simulation studies were performed to see if skewness of the regression function influenced the distributions of the examined regression estimators. The situations used in the simulations were

$$\mu(x) = -cx^2 + dx^3 \qquad \text{for } -10 \leq x \leq 10$$

and

$$\mu(x) = \begin{cases} cx & \text{for } x \leq 0 \\ -dx & \text{for } x > 0 \end{cases} \qquad \text{for } -25 \leq x \leq 25$$

where $Y_i = \mu(x_i) + \varepsilon_i$ and $\varepsilon_i \sim N(0;1)$. The x-observations were chosen equidistantly. The position of the nearest x-value was uniformly distributed in the interval $(-0.5\Delta, 0.5\Delta)$ centered at x_{max}. The result from 25 000 replications indicated that there was no essential influence by a small skewness on the distributions of any of the regression estimators $\hat{Y} \mid x_i$, $\hat{Y}_{max}$ and $\hat{X}_{max}$.

When d increased compared to c the distributions of the estimators $\hat{Y} \mid x_i$ and $\hat{Y}_{max}$ were still not influenced in a fundamental way. However for $\hat{X}_{max}$ we obtained a difference in the distributions especially in the standardized kurtosis. The results from the simulation study with piecewise linear functions are shown in table 7.8.

TABLE 7.8. $\frac{\mu_4}{\sigma^4}$

	d		
c	**0.50**	**1.00**	**1.50**
0.5	3.970	4.678	5.238
1.0	4.633	3.862	3.859
1.5	5.214	3.890	3.453
2.0	5.606	4.104	3.329
3.0	6.190	4.514	3.410

Table 7.8. Random numbers were generated so that $Y \sim N(\mu(x);1)$ for 51 equidistant x-observations in the x-interval $-25 \leq x \leq 25$, where $\mu(x)=cx$ for $x \leq 0$ and $\mu(x)=-dx$ for $x>0$. The standardized 4:th moments of $\hat{X}_{max}$ were estimated for $0.5 \leq c \leq 3.0$ and $0.5 \leq d \leq 1.5$.

First we can conclude that the values of the standardized skewness are symmetric around c=d. From the simulation results we can also see that for d=0.5 the standardized kurtosis increases when c increases. For d=1.0 and d=1.5 the standardized kurtosis first seems to decrease to a minimum and then increase. For different values of c we obtain the corresponding results. If c=0.5 then the standardized kurtosis increases when d increases. For the other examined values of c it is the other way round.

7.3. The estimation of the distribution of the inflection point in sigmoid regression for stochastic X_i.

For sigmoid regression we have examined the problem of estimation of the inflection point (x_{inf}, y_{inf}) in the chapters concerning bias and variance properties. The estimate distributions of both x_{inf} and

y_{inf} obtained by the estimation method are examined by simulations.

7.3.1. The distribution of the estimated y-value of the inflection point.

The distribution of the estimated y-value of the inflection point was examined by simulations according to the following:

Suppose we have a regression function $y_i = \mu(x) + \varepsilon_i$. Let $\mu(x) = E(Y|x)$ be a sigmoid regression function with the inflection point in x = 0 and $\varepsilon_i \sim N(0, \sigma(x_i)^2)$, where $\sigma(x_i)^2$ are bounded. The simulation study was performed for the symmetric regression functions

$$\mu_1(x) = a + 10 * \frac{e^x}{1 + e^x}$$

$$\mu_2(x) = a + 10 * \Phi(x)$$

and the non-symmetric regression function

$$\mu_3(x) = a + 10 * (1 - e^{-e^x}).$$

In all situations we used equidistant x-observations in the interval $-10 \le x \le 10$. The number of x-observations, N, varied between 11 and 31 in the examined interval. The position of the nearest x-value was uniformly distributed in the interval $(-0.5\Delta, 0.5\Delta)$ centered at x_{inf}. From the simulations using 25 000 replications the 3:rd, μ_3, and the 4:th, μ_4, moments of $\hat{Y}_{inf}$ were estimated. The standardized skewness μ_3/s^3 and the standardized kurtosis μ_4/s^4 were determined.

Since the simulation situation was symmetric around y_{inf}, μ_3/σ^3 = 0.0 for the two symmetric regression functions. The results for the non-symmetric regression function is shown in table 7.9.

TABLE 7.9.

N	11	16	21	26	31
$\frac{\mu_3}{\sigma^3}$	-0.350	-0.225	-0.222	-0.227	-0.227

Table 7.9. Random numbers were generated so that Y ~ N(μ(x);1) for equidistant x-observations in the x-interval -10 ≤ x ≤ 10, where μ(x) was an extreme value function. The standardized 3:rd moments of $\hat{Y}_{inf}$ were estimated for 11 ≤ N ≤ 31.

From this table we can see that the standardized skewness increases when the number of x-observations, N, increases from 11 to 21 in the examined interval. When N gets bigger than 21 the standardized skewness decreases again. From the same simulation study we also obtained the standardized kurtosis, μ_4/σ^4, for the three different regression functions. The results from 25 000 replications are shown in table 7.10.

From this table we see that the normalized kurtosis decreases for the symmetric regression functions when the number, N, of x-observations in the examined x-interval increases. The normalized kurtosis of the non-symmetric regression function decreases to a minimum and then increases again. It also seems that μ_4/s^4 of the non-symmetric regression function is smaller than the

TABLE 7.10.	$\mu_1(x)$	$\mu_2(x)$	$\mu_3(x)$
Number of x-observations N	$\frac{\mu_4}{s^4}$	$\frac{\mu_4}{s^4}$	$\frac{\mu_4}{s^4}$
11	2.813	2.183	2.726
16	2.218	2.032	2.129
21	2.129	2.030	2.090
26	2.101	2.023	2.097
31	2.098	2.012	2.099

Table 7.10. Random numbers were generated so that $Y \sim N(\mu_i(x);1)$ for equidistant x-observations in the x-interval $-10 \le x \le 10$, where $\mu_1(x)$ was a logitstic function, $\mu_2(x)$ was the probit function and $\mu_1(x)$ was an extreme-value function. The standardized 4:th moments of $\hat{Y}_{inf}$ were estimated for these three situations for $11 \le N \le 31$.

kurtosis of the logit function. The smallest normalized skewness was obtained for the cumulative Gaussian function.

7.3.2. The distribution of the estimated x-value of the inflection point.

Another result obtained from the preceding simulation study was what shape of the distribution of the estimated x_{inf} we would obtain by the estimation method.

In the simulation study 25 000 replications were used for the same three different regression functions as in the preceding simulation study. We also used the same assumptions, namely $Y_i = \mu(x) + \varepsilon_i$, where $\varepsilon_i \sim N(0;1)$ for $-10 \le x \le 10$. The x-observations were chosen equidistantly. The position of the nearest x-value was uniformly distributed in the interval (-0.5Δ, 0.5Δ) centered at x_{inf}. Since the simulation situation was symmetric around x_{inf}, $\mu_3/\sigma^3 = 0.00$ for the two symmetric regression functions. The simulation results for the non-symmetric regression function indicated as expected that the standardized skewness for this situation also was 0.00.

From the same simulation study we also obtained the standardized kurtosis, μ_4/σ^4, for the three different regression functions. The results from 25 000 replications are shown in table 7.11.

TABLE 7.11.	$\mu_1(x)$	$\mu_2(x)$	$\mu_3(x)$
Number of x-observations N	$\frac{\mu_4}{s^4}$	$\frac{\mu_4}{s^4}$	$\frac{\mu_4}{s^4}$
11	6.622	5.851	8.198
16	4.447	3.188	3.988
21	3.274	3.027	3.441
26	2.996	2.920	3.388
31	2.910	2.863	3.268

Table 7.11. Random numbers were generated so that $Y \sim N(\mu_i(x);1)$ for equidistant x-observations in the x-interval $-10 \leq x \leq 10$, where $\mu_1(x)$ was a logitstic function, $\mu_2(x)$ was the probit function and $\mu_1(x)$ was an extreme value function. The standardized 4:th moments of $\hat{X}_{inf}$ were estimated for these three situations for $11 \leq N \leq 31$.

From this table we see that the normalized kurtosis behaves in the same way as the normalized kurtorsis for the estimated y_{inf}, that is the kurtosis decreases when the number, N, of x-observations in the examined x-interval increases. We can also see that the standardized kurtosis is smaller for the symmetric regression functions than for the non-symmetric.

8. CONFIDENCE INTERVALS.

In some applications it is desirable to be able to draw an inference about an unknown population quantity. In such situations the calculation of confidence limits is appropriate.
In this chapter we will propose how to use the knowledge of bias, variance and distribution to construct approximate confidence intervals. The results from chapter 5-7 will be used in this discussion. In the simulations we have only treated regression functions of the form $y_k = \mu(x_k) + \varepsilon_k$, k=1,....,N, where $\varepsilon_k \sim N(0, 1)$ and the x-observations have been equidistant with the distance Δ between two successive x-observations. The position of the nearest x-value was uniformly distributed in the interval $(-0.5\Delta, 0.5\Delta)$ centered at x=0.0. Therefore this will also be the situation that is discussed here.
In chapter 7 we have examined the distributions of some estimates under these assumptions. It is of course interesting to see in which situations we can approximate the distributions by the normal distribution. Therefore we compare the obtained simulation results of the standardized skewness, μ_3/σ^3, and the standardized kurtosis, μ_4/σ^4, to the corresponding values of the normal distribution, which are 0.0 and 3.0.
Using these criteria we see that in general the distributions of $\hat{X}_{max}$ for regression functions with constant curvature, regression functions with piecewise linear functions and distributions with a

small skewness could be approximated by the normal distribution. This is also the case for $\hat{Y}_{inf}$.

To be able to calculate a confidence interval we need an estimate of the variance. This is discussed in chapter 6.

Under these assumptions we can construct approximate confidence intervals for each of $\hat{X}_{max}$ and $\hat{Y}_{inf}$ in the same way as for any normally distributed variables. When we have a situation where this approximation is not appropriate, we can use an Edgeworth expansion described in chapter 7:

Suppose we have a random variable X with a corresponding expected value, $\mu(X)$, and standard variation, $\sqrt{\hat{Var}(X)}$. The distribution of the normalized variable,

$$U = \frac{\hat{\mu}(X) - \mu(X)}{\sqrt{\hat{Var}(X)}}$$

can be described by this expansion

$$\hat{F}_{Edg}(U) = \Phi(U) - \phi(U) \left[\frac{1}{6} \left(\frac{\mu_3}{\sigma^3} \right) H_2(U) + \frac{1}{24} \left(\frac{\mu_4}{\sigma^4} - 3 \right) H_3(U) \right.$$

$$\left. + \frac{1}{72} \left(\frac{\mu_3}{\sigma^3} \right)^2 H_5(U) \right] + r_n$$

where

$$r_n = o(n^{-1/2}),$$

$\Phi(U)$ = the normal cumulative probability distribution,

$$\phi(U) = \frac{1}{\sqrt{2\Pi}} \operatorname{Exp} \{ -\frac{1}{2} U^2 \}$$

$H_2(U) = U^2 - 1$,

$H_3(U) = U^3 - 3U,$

$H_5(U) = U^5 - 10U^3 + 15U.$

But first we need some approximation of the normalized curvature, NC. As a rough estimate we can use the definition of NC in the neighbourhood of the estimated maximum point, using the estimated maximum point, $(\hat{X}_{max}, \hat{Y}_{max})$, together with the two neighbouring estimated points, $(\hat{X}_{max+1}, \hat{Y}_{max+1})$ and $(\hat{X}_{max-1}, \hat{Y}_{max-1})$, according to

$$\hat{NC} = \frac{(\hat{Y}_{max+1}-\hat{Y}_{max})(\hat{X}_{max}-\hat{X}_{max-1}) - (\hat{Y}_{max}-\hat{Y}_{max-1})(\hat{X}_{max+1}-\hat{X}_{max})}{\frac{\sigma}{\sqrt{n}}(\hat{X}_{max}-\hat{X}_{max-1})(\hat{X}_{max+1}-\hat{X}_{max})(\hat{X}_{max+1}-\hat{X}_{max-1})}.$$

A simulation study was made to examine bias- and variance-properties of the estimate of NC. The same simulation situations as in the preceding chapters 5-7 were used, ie a sample $y_k=-cx_k^2+\varepsilon_k$, $k=1,....,N$, was obtained, for the x-interval $-10 \leq x \leq 10$ where $\varepsilon_k \sim N(0, 1)$. The x-observations were chosen equidistantly with the distance 'one' between two successive x-observations. The position of the nearest x-value compared to $x_{max}=0.0$ was uniformly distributed in the interval (-0.5, 0.5). The normalized curvature was estimated using 500 replications. Bias and variance of the estimates were obtained. The results are shown in table 8.1. From the table we can see that bias increases up to a maximum obtained around c=1 and then decreases again. This means that we overestimate the curvature when we use the proposed estimation method. However the estimate is rather close to the true value.

TABLE 8.1.

c	$\hat{NC}$	$\hat{Bias}(\hat{NC})$	$\hat{Var}(\hat{NC})$
0.2	0.667	0.447	0.0861
0.5	1.144	0.644	0.2227
0.7	1.389	0.689	0.3105
0.9	1.606	0.706	0.4213
1.0	1.735	0.735	0.4808
1.1	1.811	0.711	0.5379
1.3	2.008	0.708	0.6355
1.5	2.207	0.707	0.7670
2.0	2.598	0.598	1.0032
2.5	3.008	0.508	1.1880

Table 8.1. Random numbers were generated so that $Y \sim N(-cx^2, 1)$ in the x-interval $-10 \leq x \leq 10$. The normalized curvature, NC, was estimated using 500 replications. Bias and variance of $\hat{NC}$ were estimated.

In the table we can also see that the estimated variance increases when c increases. In figure 8.1 we have illustrated the relation between $\hat{NC}$ and c.

This means that using the estimate of NC we can use the obtained simulation results for bias, the variance, skewness and kurtosis and substitute them in the formulas according to

$$P(\hat{F}_{Edg}(U_{0.025}) \leq U = \frac{\hat{\mu}(X) - \mu(X)}{\sqrt{\hat{Var}(X)}} \leq \hat{F}_{Edg}(U_{0.975})) = 0.95.$$

This can be evaluated to

$$P[\hat{\mu}(X) - \hat{F}_{Edg}(U_{0.975}) \sqrt{\hat{Var}(X)} \leq \mu(X) \leq \hat{\mu}(X) + \hat{F}_{Edg}(U_{0.025}) \sqrt{\hat{Var}(X)}].$$

Suppose that we have used our proposed estimation method in some situations and have obtained an estimate, $\hat{Y}(X)$. A rough

FIGURE 8.1.

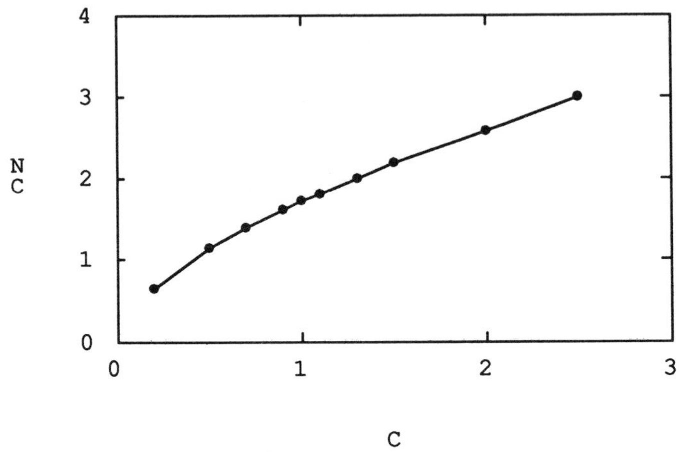

Figure 8.1. Random numbers were generated such that $Y \sim N(-cx^2, 1)$ in the x-interval $-10 \leq x \leq 10$. The normalized curvature, NC, was estimated using 500 replications. The relation between c and $\hat{NC}$ was estimated.

estimate of NC gives us the obtained simulation results for bias, the variance, skewness and kurtosis.

If we have a bias, B, we use the estimate $\hat{\mu}(X) = \hat{Y}(X) - B$. Then we normalize the variable and obtain the described situation where we can calculate the approximate 95% confidence interval.

9. A BIOLOGICAL PROBLEM.

In the earlier parts of this thesis we have used normally distributed variables, $Y_k = \mu(x_k) + \varepsilon_k$, k=1,....,N, where $\varepsilon_k \sim N(0, \sigma(x_k)^2)$. We are of course interested in the possibility to generalize the obtained results. For distributions that belong to the exponential family it is sometimes possible to show that convex/concave regression estimates are in fact maximum likelihood estimates. Since one of the most important applications is found in bioassay, we here restrict ourselves to find the convex/concave ML regression estimates for the parameters of the binomial distribution. We will also discuss the special problem that arises when we combine the convex and the concave phases to sigmoid regression.

A common problem in bioassay is the problem of estimating an S-shaped form of binomial parameters. We could for instance be interested in the death rate at some doses of drugs. This means that we have ordered levels of drugs $x_1 \le x_2 \le ....\le x_N$, that correspond to death probabilities $p(x_1) \le p(x_2) \le .... \le p(x_N)$, which forms a sigmoid curve.

Suppose that for each x_i we have n_i independent trials with corresponding 0,1 responses, Y_{ij}. A common way of estimating $p(x_i)$ is to use the sufficient statistics $\bar{y}_i$ for each x_i. But sometimes $\bar{y}_i$, $i=1,...,N$, does not fulfill the restrictions for the form of the dose-response curve. In these situations we need a non-trivial estimate of the function. The S-shaped function consists of one convex phase and one concave phase. To be able to deal with sigmoid curves we start by examining convex functions. Since MLE have nice statistical properties we try to prove that we obtain the MLE of the regression function using our proposed estimation procedure directly.

Suppose that there is a convex function, Φ, which is finite in an interval I. This means that Φ is continuous and differentiable from both left and right at each point within I. Denote a derivate of Φ at x by $\phi(x)$. If the right and the left derivate do not coincide, let $\phi(x)$ be a number between the two. Suppose that $\Phi(x)$ has a derivative $\phi(x)$, which is increasing and convex. Then for each $u,v \in I$ we define a function $\Delta_\Phi(u,v)$ by

$$\Delta_\Phi(u,v) = \Phi(u) - \Phi(v) - (u-v)\phi(v).$$

From this definition is directly obtained that

$$\Delta_\Phi(r,t) = \Delta_\Phi(r,s) + \Delta_\Phi(s,t) - (r-s)(\phi(s) - \phi(t)).$$

Since Φ is strictly convex we have $\Delta_\Phi(r,t) \geq 0$. Further if $f(x)$ is a convex function and Φ is increasing and convex then $\Phi(f(x))$ is convex. This is easily shown using the second derivate of $\Phi(f(x))$

$$\frac{d''\Phi(f)}{dx} = \Phi''(f(x))\ (f'(x))^2 + f''(x)\ \Phi'(f(x))$$

which is positive. Using these notations we can formulate the following theorem:

IS THIS RESULT TRUE?: Suppose that the interval I contains the range of the function g and that $\Phi(x)$ is increasing and convex. Let Δ_Φ be defined in the same way as above. If a function f is convex and if the range of f is in the interval I then

$$\sum_{x\in X}\Delta_\Phi(g(x),f(x))w(x) \geq \sum_{x\in X}\Delta_\Phi(g(x),g^*(x))w(x) + \sum_{x\in X}\Delta_\Phi(g^*(x),f(x))w(x),$$

where the increasing and convex estimate $g^*(x)$ minimizes $\sum_{x\in X}\Delta_\Phi\big(g(x),f(x)\big)w(x)$ in the class of all increasing and convex functions f with range in I. Then $g^*(x)$ maximizes

$$\sum_{x\in X}\Big\{\Phi(f(x)) + \big[g(x) - f(x)\big]\ \phi(f(x))\Big\}w(x).$$

The function $g^*(x)$ is unique if Φ is increasing and convex.

According to Robertson et al (1989) a similar result holds for isotonic functions f(x). They showed that an inequality $\sum_{x\in X}[g(x)-g^*(x)]\phi[f(x)]w(x) \leq 0$ holds for isotonic f(x). The gap in a proof of the above result which we have not been able to fill out consists in showing that $\sum_{x\in X}[g(x)-g^*(x)]\phi[f(x)]w(x) \leq 0$ for convex f(x). However, we give a hint of a proof:

Use the relation $\Delta_\Phi(r,t) = \Delta_\Phi(r,s) + \Delta_\Phi(s,t) - (r-s)(\phi(s) - \phi(t))$ above with r=g(x), s= $g^*(x)$ and f=f(x). Then we get the difference between the left side and the right side of the inequality in the theorem. This difference can be written as

$$\sum_{x\in X}[g(x) - g^*(x)]\left(\phi[g^*(x)] - \phi[f(x)]\right)w(x) \geq 0.$$

1) $\sum_{x\in X}[g(x) - g^*(x)]\,\phi[g^*(x)]w(x) = 0$. This is warranted by the fact that if the value of the sum is positive ie $\sum_{x\in X}[g(x) - g^*(x)]\,\phi[g^*(x)]w(x) > 0$ then a function $g^*(x) + \varepsilon\phi[g^*(x)]$ would give a smaller m.s. error than $g^*(x)$ and still be increasing and convex. The case with a negative sum is analogous.

2) A suitable detail that we failed to prove is that since ϕ is increasing and convex then this implies that $\sum_{x\in X}[g(x)-g^*(x)]\phi(f(x))w(x) \leq 0$. As said before, this inequality holds for isotonic functions f(x), but we haven't been able to extend it to convex functions. However,

from 1) and 2) we obtain the proof of the first part of the theorem.

The conclusion that $g^*(x)$ minimizes the equation $\sum \Delta_\Phi[g(x),f(x)]w(x)$ for $x \in X$ follows directly from the fact that $\Delta_\Phi \geq 0$. It is easily seen that $g^*(x)$ maximizes the equation $\sum \{\Phi[f(x)] + [g(x)-f(x)]\phi[f(x)]\}w(x)$ for $x \in X$ by rewriting $\Delta_\Phi[g(x),f(x)]$ as $\Phi[g(x)] - \Phi[f(x)] - [g(x)-f(x)]\phi[f(x)]$, where $\Phi[g(x)]$ does not depend on f(x). Finally we can conclude that the minimizing/maximizing functions are unique since the term $\sum_{x \in X} \Delta_\Phi(g^*(x),f(x))w(x) > 0$ unless $g^*(x) = f(x)$.

Let f denote an arbitrary function on X whose values are bounded between 0 and 1. Writing out $\sum \Delta_\Phi[\bar{y}(x),f(x)]w(x)$, $x \in X$, with $\Phi(z) = z \ln[z] + (1-z)\ln[1-z]$, and omitting all terms which do not depend on f(x) we obtain

$$\sum_{x \in X} \Delta_\Phi[\bar{y}(x),f(x)]w(x) = \sum_{x \in X} \{\bar{y}(x)\ln[f(x)] + (1-\bar{y}(x))\ln[1-f(x)]\}w(x).$$

If the above result could be proved the isotonic regression $\bar{y}^*$ of $\bar{y}$ with weights w(x) provides the MLE of p(x) subject to the assumption that the estimates must form an isotonic curve.

Since we are interested in the possibilities to use least squares sigmoid regression in bioassay, this theorem can give us an idea of how to obtain the MLE of $\mu(x)$, where $\mu(x)$ is a sigmoid function of binomial parameters.

Let n_i be the number of y-observations at x_i. Denote the number of times an event happens at x_i by m_i. The distribution of the relative frequencies $\bar{Y}_i$ is

$$P(\bar{Y}_i=y) = \binom{n_i}{n_i y} p_i^{n_i y} (1-p_i)^{n_i(1-y)}, \text{ where } y=0, \frac{1}{n_i}, \frac{2}{n_i}, \ldots, 1.$$

and corresponding log likelihood equation equals

$$\ln\left[P(\bar{Y}_i=y_i)\right] = \ln\binom{n_i}{n_i y_i} + n_i y_i \ln\left[\frac{p_i}{1-p_i}\right] + n_i \ln\left[1-p_i\right].$$

which equals the expression above. The following expression is to be maximized

$$\sum_{x\in X}\{\Phi(f(x)) + [g(x)-f(x)]\phi(f(x))\}w(x)$$

where $w(x)=n_i$, $g(x)=y_i=\frac{m_i}{n_i}$ and $f(x)=p_i$.

According to McCullagh and Nelder (1983) it is possible to use an iterative procedure for general linear models in a similar situation. The process is iterative because the weight **W** depends on the fitted values for which only current estimates are available. The procedure underlying the iteration is as follows. Let $\tilde{\mathbf{p}}_0$ be the current estimates. Form the adjusted weights $\tilde{\mathbf{W}}_0$ according to

$$\tilde{\mathbf{W}}_0^{-1} = \frac{\delta^2 \ln[P_p(Y_i=p_i)]}{\delta p_i^2}.$$

Now regress $\tilde{\mathbf{p}}$ on x_i with weights $\tilde{\mathbf{W}}_0$ to obtain new estimates $\tilde{\mathbf{p}}_1$. Form the adjusted weights $\tilde{\mathbf{W}}_1$. Repeat until the estimates converges to limit values. When you have found $\tilde{p}_i$, i=1,..,N, so that $\tilde{p}_i$ minimize $\sum_{i=1}^{N}\tilde{w}_i(\tilde{p}_i-\hat{p}_i)^2$, then $\tilde{p}_i$, i=1,..,N, is a local optimum, ie for any a_i, i=1,...,N, and for Δ small enough the following holds

$$\textbf{log lik}(y_1, y_2,...,y_N, \tilde{p}_1, \tilde{p}_2,...,\tilde{p}_N) \geq \textbf{log lik}(y_1, y_2,...,y_N, \tilde{p}_1+a_1\Delta, \tilde{p}_2+a_2\Delta,..., \tilde{p}_N+a_N\Delta).$$

Since the estimates obtained by sigmoid regression consist of piecewise linear functions this iterative procedure is applicable to our proposed estimation method according to the following: Denote the sigmoid regression estimates of p_i by $\tilde{p}_i$. Then the proposed next set of weights, $\tilde{w}_i$, are the second derivatives of the log-likelihood, $\delta^2 \ln[P_p(\bar{Y}_i=\hat{p}_i)]/\delta p_i^2$, ie $\tilde{w}_i=n_i[(1-\hat{p}_i)/(1-\tilde{p}_i)^2+(\hat{p}_i)/\tilde{p}_i^2]$, where $\hat{p}_i$ equal the starting values in the iterations. Use sigmoid regression on $\hat{p}_i$, i=1,..,N, with the corresponding weights. Since $\tilde{p}_i \approx \hat{p}_i$ we can start the iteration procedure using the approximated weights $n_i/[(1-\hat{p}_i)\hat{p}_i]$. Now we obtain a new estimate $\tilde{p}_i^{(1)}$, i=1,...,N, of the sigmoid regression function. From these we can calculate new weights

$$\tilde{w}_i^{(1)} = n_i\left(\frac{1-\hat{p}_i}{(1-\tilde{p}_i^{(1)})^2}+\frac{\hat{p}_i}{(\tilde{p}_i^{(1)})^2}\right).$$

We use sigmoid regression on the new estimates $\tilde{p}_i^{(1)}$, i=1,...,N, with corresponding weights $\tilde{w}_i^{(1)}$ and obtain a new estimate $\tilde{p}_i^{(2)}$, i=1,...,N, of the sigmoid regression function. From these we can calculate new weights $\tilde{w}_i^{(2)}$. Repeat until changes in $\tilde{w}_i^{(k)}$ are sufficiently small. A problem arises when $\hat{p}_i$ or $\tilde{p}_i$ is 0.0 or 1.0. In these situations we can't use the usual formula to obtain the weight but have to give the weight a high value. When this situation occurs we choose to use the weight at least ten times as big as the largest of the other weights. Another problem is that the algorithm is sensitive to starting values, since the iterations only find a local maximum. After examining a few alternatives we found that, if $\hat{p}_i$ equals the monotonized m_i/n_i, we obtained starting values that gave us the largest likelihood in all the examined situations. A simulation study was performed to illustrate the iterative procedure.

Suppose we have a sample $y_i = \pi(x_i) + \varepsilon_i$, i=1,....,N, where $\pi(x_i)$ is a sigmoid regression function with equidistant x-observations in the interval $-d \le x \le d$ and $\varepsilon_i \sim Bin(n(x_i); \pi(x_i))$. The number of x-observations was 11 and $n(x_i)=10$ for each x_i.The examined regression function was of the type $\pi(x) = e^x/(1 + e^x)$ in the interval $-2.5 \le x \le 2.5$. Ten replicates were examined. All cases converged to the MLE within 15 iterations. Two of the examined cases are shown in table 9.1.

TABLE 9.1. 1:st replicate / 2:nd replicate

x_i	1:st replicate $\frac{m_i}{n_i}$	$\hat{p}_i$	$\tilde{p}_i$	$\tilde{w}_i$	2:nd replicate $\frac{m_i}{n_i}$	$\hat{p}_i$	$\tilde{p}_i$	$\tilde{w}_i$
-2.5	0.1	0.100	0.1212	0.1192	0.0	0.0	0.0000	10.0000
-2.0	0.4	0.107	0.1419	0.1410	0.0	0.0	0.0011	0.0170
-1.5	0.0	0.167	0.1626	0.1122	0.1	0.1	0.1089	0.1530
-1.0	0.1	0.167	0.1833	0.0930	0.2	0.2	0.1866	0.1112
-0.5	0.2	0.200	0.2133	0.0850	0.3	0.3	0.2643	0.0894
0.0	0.4	0.400	0.3661	0.0670	0.3	0.3	0.3420	0.0669
0.5	0.5	0.500	0.5189	0.0601	0.6	0.6	0.6000	0.0666
1.0	0.8	0.800	0.7789	0.0809	0.8	0.8	0.8000	0.0999
1.5	0.8	0.800	0.8057	0.0976	1.0	0.9	0.8918	0.1546
2.0	0.8	0.800	0.8324	0.1237	0.8	0.9	0.9115	0.2213
2.5	1.0	1.000	0.8591	0.0203	1.0	1.0	0.9312	0.0184
	8 iterations SSE=0.04662				6 iterations SSE=0.01040			

Table 9.1. An iterative procedure was used to obtain the MLE, $\tilde{p}_i$, with corresponding weights, $\tilde{w}_i$, for two replicates. The number of iteration before the procedure converged was noticed in each case and the corresponding SSE was calculated.

From the table we can see that $\tilde{w}_i$ are bigger at the ends of the x-interval than in the middle except when $\hat{p}_i$ is either 0.00 or 1.00. In the second replicate we observe that if the sigmoid estimate is 0.0 then we have chosen the corresponding weight to be big, 10.0, compared to the largest of the other weights, 0.2213. The same is applied if the sigmoid estimate is 1.0. If instead of 10.0 we choose the weights corresponding to the sigmoid estimates $\tilde{p}_i$ =0.0 or $\tilde{p}_i$ =1.0 equal 100.0, then neither the weights corresponding to the other estimates nor the sigmoid estimates changed in the examined situations.

Since the iteration procedure demands much time and resource we have examined alternative procedures that don't need so much effort. The first alternative was to use sigmoid regression on the observed values m_i/n_i together with the approximate weights using the assumption that $\tilde{p}_i \approx \hat{p}_i$, ie $\tilde{w}_i = n_i/[(1-\hat{p}_i)\,\hat{p}_i]$, i=1,...,N. Denote these weights by $\tilde{w}_i^{(1)}$ and the corresponding sigmoid estimates by $\tilde{p}_i^{(1)}$. The second alternative was to use sigmoid regression on the observed proportions together with the weights $\tilde{w}_i$=1.0, i=1,...,N. Denote the weights in this second alternative by $\tilde{w}_i^{(2)}$ and the corresponding sigmoid estimates by $\tilde{p}_i^{(2)}$.

Sigmoid regression of m_i/n_i on x_i was used on the same replicates as for the iteration procedure. Each replicate was examined for both $\tilde{w}_i^{(1)}$ and $\tilde{w}_i^{(2)}$. The results are shown in table 9.2.

From this table we can see that using the alternative weights $\tilde{w}_i^{(2)}$ = 1.0, i=1,..,N, gives us an estimate that is rather close to the MLE. This is the reason why we use this simple alternative in our simulations instead of the iteration procedure.

TABLE 9.2.

1:st replicate

x_i	$\frac{m_i}{n_i}$	$\tilde{p}_i$	$\tilde{w}_i$	$\tilde{p}_i^{(1)}$	$\tilde{w}_i^{(1)}$	$\tilde{p}_i^{(2)}$	$\tilde{w}_i^{(2)}$
-2.5	0.1	0.1212	0.1192	0.0027	1.1111	0.1500	1.0000
-2.0	0.4	0.1419	0.1410	0.0027	0.4167	0.1500	1.0000
-1.5	0.0	0.1626	0.1122	0.0027	10.0000	0.1500	1.0000
-1.0	0.1	0.1833	0.0930	0.1000	1.1111	0.1500	1.0000
-0.5	0.2	0.2133	0.0850	0.2117	0.6250	0.1962	1.0000
0.0	0.4	0.3661	0.0670	0.3650	0.4167	0.3769	1.0000
0.5	0.5	0.5189	0.0601	0.5182	0.4000	0.5577	1.0000
1.0	0.8	0.7789	0.0809	0.7430	0.6250	0.7385	1.0000
1.5	0.8	0.8057	0.0976	0.8285	0.6250	0.8077	1.0000
2.0	0.8	0.8324	0.1237	0.9140	0.6250	0.8769	1.0000
2.5	1.0	0.8591	0.0203	0.9995	10.0000	0.9462	1.0000
		SSE=0.04662		SSE=0.0884		SSE=0.10360	

2:st replicate

x_i	$\frac{m_i}{n_i}$	$\tilde{p}_i$	$\tilde{w}_i$	$\tilde{p}_i^{(1)}$	$\tilde{w}_i^{(1)}$	$\tilde{p}_i^{(2)}$	$\tilde{w}_i^{(2)}$
-2.5	0.0	0.0000	10.0000	0.0000	10.0000	0.0000	1.0000
-2.0	0.0	0.0311	0.0170	0.0002	10.0000	0.0200	1.0000
-1.5	0.1	0.1089	0.1530	0.0009	1.1111	0.1000	1.0000
-1.0	0.2	0.1866	0.1112	0.1755	0.6250	0.1800	1.0000
-0.5	0.3	0.2643	0.0894	0.2631	0.4762	0.2600	1.0000
0.0	0.3	0.3420	0.0669	0.3508	0.4762	0.3400	1.0000
0.5	0.6	0.6000	0.0666	0.6000	0.4167	0.6000	1.0000
1.0	0.8	0.8000	0.0999	0.8000	0.6250	0.8000	1.0000
1.5	1.0	0.8918	0.1546	0.9994	10.0000	0.9333	1.0000
2.0	0.8	0.9115	0.2213	0.9994	0.6250	0.9333	1.0000
2.5	1.0	0.9312	0.0184	0.9994	10.0000	0.9333	1.0000
		SSE=0.01040		SSE=0.02730		SSE=0.03067	

Table 9.2. An iterative procedure was used to obtain the MLE, $\tilde{p}_i$, with corresponding weights, $\tilde{w}_i$. Two alternative weights, $\tilde{w}_i^{(1)} = n_i[(1-\hat{p}_i)\,\hat{p}_i]^{-1}$ with corresponding sigmoid estimate $\tilde{p}_i^{(1)}$ and $\tilde{w}_i^{(2)} = 1.0$ with corresponding sigmoid estimate $\tilde{p}_i^{(2)}$ were determined. Sums of squares of errors were calculated in each separate case.

9.1. Simulation results in some different situations using binomially distributed variables with equal weights.

To confirm the possibility to generalize the obtained simulation results three simulation studies were performed. Since the estimates using $w_i=1$, $i=1,...,N$, are very close to the ML estimates obtained from the described iteration procedure, we have used this simple situation with equal weights in all the simulations. In all three studies the nature of bias and variance for binominally distributed variables have been examined.
Since a common problem in bioassay is to estimate sigmoid curves the examined regression functions were s-shaped in these situations.

In the first study we used a logistic regression function. Sigmoid regression was compared to regression using the correct assumption that the regression function was logistic. Comparison was also made to regression using the false assumption that the regression function was an extreme-value function, which is a skew sigmoid curve.

In the second simulation study we used an extreme-value regression function. In this study we compared sigmoid regression to the same two alternatives as in the preceding study. But now the assumption of a logistic regression function was false and the assumption of an extreme-value regression function was correct.

The third simulation study was performed to examine the difference in bias and variance between sigmoid regression and probit regression when the regression function was a probit function.

9.1.1. Comparison of bias obtained for sigmoid regression to bias obtained in some symmetric and one non-symmetric situations.

Suppose we have a sample $y(x_k)$ where $Y(x_k) \sim Bin(n(x_k), \pi(x_k))$ and $\pi(x_k)$ is a sigmoid function with equidistant fixed x-observations in the interval $-d \le x \le d$. The number of x-observations was 21 and $n(x_k)=10$ for each x_k. In the first simulation study the regression function was of the form $\pi(x) = (e^x/(1 + e^x)$ in the interval $-5 \le x \le 5$. The result from 10 replications, where the same random numbers were used in sigmoid regression as as well logistic and extreme value regression, shows $\hat{y}_{bias}$ in table 9.3.

TABLE 9.3. Bias

x	sigmoid regression	logistic regression	extreme value regression
-5.00	0.000	0.000	0.014
-4.50	-0.001	-0.001	0.017
-4.00	-0.003	-0.001	0.025
-3.50	-0.005	-0.001	0.025
-3.00	-0.001	-0.001	0.028
-2.50	0.008	-0.001	0.020
-2.00	0.022	-0.001	0.025
-1.50	0.007	-0.001	0.016
-1.00	-0.017	-0.001	0.000
-0.50	-0.025	-0.002	-0.019
0.00	-0.021	-0.004	-0.035
0.50	0.024	-0.004	-0.029
1.00	0.014	-0.004	-0.016
1.50	-0.003	-0.003	0.013
2.00	-0.015	-0.001	0.037
2.50	-0.008	-0.001	0.044
3.00	-0.003	0.000	0.037
3.50	-0.003	0.001	0.027
4.00	-0.004	0.001	0.018
4.50	-0.003	0.001	0.011
5.00	-0.003	0.000	0.007

Table 9.3. Binomial distributed random numbers were generated around the logistic function. These random numbers were used in sigmoid regression, logistic regression and extreme-value regression. Bias in these three situations were determined using 10 replications.

From this table we can see that logistic regression gave the smallest bias for almost all x-values. This result was expected since the estimation usually gets better if the regression function is known. In the situation when we made the false assumption that the regression function was an extreme-value function, we obtained the largest bias for almost all x-values. In this situation bias was small for $x < 0$ and increased for $x > 0$. This result is logical since the extreme-value function is skew and the correct regression function is symmetric. When we used sigmoid regression we obtained a bias that was bigger than for the logistic function but smaller than for the extreme value function for almost all x-values.
In the second simulation situation sigmoid regression was compared to logistic regression and to regression under the assumption that the regression function was an extreme-value function as in the preceding study. But in this study the regression function was of the form $\pi(x) = 1- [\exp(-\exp(x))]$ in the interval $-2.5 \le x \le 2.5$. The result from 10 replications, where the same random numbers were used in sigmoid regression as well as logistic and extreme-value regression is shown in table 9.4. From the results in this table we can see that the obtained bias will be small if the assumption that the regression function is an extreme value function is correct. If on the other hand we made the false assumption that the regression function was logistic, then bias was bigger for almost all x-observations. For sigmoid regression we obtained a bias that was bigger than when the regression function was known and smaller than for logistic regression for almost all x-values. The conclusion from these two simulation studies is that if we don´t know the shape of the regression function, sigmoid regression is probably the best choice.

TABLE 9.4. Bias

x	sigmoid regression	logistic regression	extreme value regression
-2.50	0.004	0.038	0.010
-2.25	0.004	0.043	0.013
-2.00	0.009	0.048	0.016
-1.75	0.019	0.050	0.020
-1.50	0.032	0.047	0.021
-1.25	0.049	0.040	0.025
-1.00	0.069	0.028	0.026
-0.75	0.019	0.009	0.024
-0.50	0.000	-0.006	0.024
-0.25	0.007	-0.026	0.012
0.00	0.001	-0.029	0.010
0.25	0.028	-0.019	0.003
0.50	0.005	-0.001	0.003
0.75	0.016	0.020	0.001
1.00	0.028	0.005	0.005
1.25	0.029	0.042	0.008
1.50	0.024	0.040	0.007
1.75	0.013	0.033	0.004
2.00	0.006	0.026	0.001
2.25	0.000	0.018	0.000
2.50	0.000	0.013	0.000

Table 9.4. Binomial distributed random numbers were generated around the extreme-value function. These random numbers were used in sigmoid regression, logistic regression and extreme-value regression. Bias in these three situations were determined using 10 replications.

To see if there are any essential differences in the conclusions using the logit or the probit functions as the regression functions another simulation study was performed. In this study we compared sigmoid regression to the probistic regression when the regression function was the cumulative standard normal distribution

$$\pi(x) = \int_{-\infty}^{x} \phi(t)\,dt \quad \text{in the interval } -2.5 \le x \le 2.5.$$

TABLE 9.5. Bias / Bias

x	sigmoid regression	probistic regression	x	sigmoid regression	probistic regression
-2.50	-0.003	0.000	0.25	0.033	0.023
-2.25	-0.004	-0.001	0.50	0.036	0.008
-2.00	-0.007	-0.003	0.75	0.014	0.011
-1.75	-0.008	-0.004	1.00	0.003	0.011
-1.50	-0.004	-0.007	1.25	-0.002	0.010
-1.25	-0.007	-0.010	1.50	-0.009	0.007
-1.00	-0.020	-0.012	1.75	-0.010	0.005
-0.75	-0.047	-0.014	2.00	-0.004	0.003
-0.50	-0.026	-0.013	2.25	0.002	0.001
-0.25	-0.005	-0.008	2.50	0.001	0.000
0.00	0.010	-0.003			

Table 9.5. Binomial distributed random numbers were generated around the probit function. These random numbers were used in sigmoid regression and probistic regression. Bias in these two situations were determined using 10 replications.

The result from 10 replications, where the same random numbers were used in sigmoid regression as well as probistic regression, shows bias obtained from the two estimation methods in table 9.5. The results are analogical to those in table 9.3 and 9.4, ie if the regression function is known, we obtain a smaller bias for almost all x-values than if it is unknown and we use sigmoid regression. But the difference is not very big. All these simulations indicate that sigmoid regression is a robust method, rather unsensitive to the form of the S-shaped function.

9.1.2. Comparison of the variance obtained using sigmoid regression to the variance obtained in some alternative situations.

As a complement to the examination of bias obtained in different situations the standard deviations for the same situations were approximated by simulations.

Suppose we have a sample $y(x_k)$ where $Y(x_k) \sim Bin(n(x_k), \pi(x_k))$ and $\pi(x_k)$ is a sigmoid curve with equidistant fixed x-observations in the interval $-d \le x \le d$. The number of x-observations was 21 and $n(x_k)=10$ for each x_k.

In the three simulation studies the regression functions were of the form

1) $\pi(x) = \dfrac{e^x}{1 + e^x}$ in the interval $-5.0 \le x \le 5.0$,

2) $\pi(x) = 1 - e^{-e^x}$ in the interval $-2.5 \le x \le 2.5$,

3) $\pi(x) = \int_{-\infty}^{x} \phi(t)\, dt$ in the interval $-2.5 \le x \le 2.5$.

The simulation studies used 10 replications each and the results in table 9.6 show the estimated standard variation of the estimates obtained from simulations using the assumption that the correct regression functions, the cumulative standard normal distribution and the extreme value function were known. We also estimated the standard deviation using sigmoid regression when the regression function was the extreme-value functions.

In table 9.6 we can see that the probistic regression gives the smallest standard deviations for almost all x-values. If we compare sigmoid regression to extreme values regression we notice that sigmoid regression gives us bigger standard deviations for almost all x-values except in the neighbourhood of the inflexion point.

For regression functions that belong to the exponential family it is easy to approximate the variance of the estimate of π_i.

TABLE 9.6. Standard deviation

x	sigmoid regression	probistic regression	extreme value regression
-2.50	0.021	0.005	0.031
-2.25	0.021	0.008	0.036
-2.00	0.031	0.013	0.041
-1.75	0.044	0.020	0.046
-1.50	0.055	0.030	0.050
-1.25	0.065	0.041	0.052
-1.00	0.069	0.052	0.051
-0.75	0.112	0.062	0.047
-0.50	0.100	0.067	0.045
-0.25	0.099	0.068	0.052
0.00	0.058	0.065	0.067
0.25	0.048	0.061	0.082
0.50	0.048	0.057	0.080
0.75	0.052	0.053	0.065
1.00	0.047	0.048	0.047
1.25	0.044	0.041	0.030
1.50	0.031	0.033	0.017
1.75	0.021	0.024	0.009
2.00	0.010	0.017	0.003
2.25	0.014	0.011	0.000
2.50	0.019	0.007	0.000

Table 9.6. Binomial distributed random numbers were generated around the probit function and the standard deviation was estimated using 10 replications. Binomial distributed random numbers were also generated around the extreme value function. These last generated random numbers were used in sigmoid regression and extreme-value regression. The standard deviation in these two situations were estimated using 10 replications.

Suppose we have a link function $g(\pi_i) = \eta_i = \sum_{i=1}^{N} x_{ij}\beta_j$. Then we obtain the variance of the estimate according to

$$\mathrm{Var}(\hat{\pi}_i) = \mathrm{Var}[g^{-1}(\sum_{i=1}^{N} x_{ij}\beta_j)] \approx (g^{-1})'[(\sum_{i=1}^{N} x_{ij}\beta_j)]\ \mathrm{Var}[\sum_{i=1}^{N} x_{ij}\beta_j] \quad \Leftrightarrow$$

$$\mathrm{Var}(\hat{\pi}) \approx (\frac{\delta\pi}{\delta\eta})^2\ \mathrm{Var}(\hat{\eta}) = (\frac{\delta\pi}{\delta\eta})^2\ \mathbf{X}\ \mathrm{Var}(\hat{\beta})\ \mathbf{X}^T.$$

For binomially distributed variables X_i such that $X_i \sim \mathrm{Bin}(n_i, \pi_i)$ for the logistic function we get the link $g(\pi) = \eta = \mathbf{ln}\,[\pi/(1-\pi)]$ with the inverse $\pi = e^{\eta}/(1+e^{\eta})$. In this situation we obtain $\mathrm{Cov}(\hat{\beta}) = (\mathbf{X}^T\mathbf{W}\mathbf{X})^{-1}$ where $\mathbf{W}=\mathrm{diag}[n_i\pi_i(1-\pi_i)]$. Thus

TABLE 9.7. Standard deviation

x	sigmoid regression	logistic regression	approximation
-5.00	0.015	0.005	0.005
-4.50	0.016	0.007	0.007
-4.00	0.019	0.010	0.010
-3.50	0.024	0.014	0.014
-3.00	0.040	0.020	0.021
-2.50	0.053	0.027	0.028
-2.00	0.067	0.034	0.036
-1.50	0.090	0.043	0.045
-1.00	0.109	0.051	0.051
-0.50	0.114	0.057	0.055
0.00	0.110	0.061	0.056
0.50	0.093	0.061	0.055
1.00	0.080	0.055	0.051
1.50	0.067	0.047	0.045
2.00	0.050	0.037	0.036
2.50	0.039	0.027	0.028
3.00	0.030	0.020	0.021
3.50	0.024	0.014	0.014
4.00	0.019	0.009	0.010
4.50	0.017	0.006	0.007
5.00	0.016	0.004	0.005

Table 9.7. Binomial distributed random numbers were generated around the logistic function. These random numbers were used in sigmoid regression and logistic regression. The standard deviation in these two situations were estimated using 10 replications. An approximation of the standard deviation was also calculated.

$$\hat{Var(\pi)} \approx \left(\frac{-e^{-\eta}}{(1+e^{-\eta})^2}\right)^2 *$$

$$\left[\frac{2\left(\sum_{i=1}^{N} x_i^2 n_i \pi_i (1-\pi_i)\right)}{\left(\sum_{i=1}^{N} x_i^2 n_i \pi_i (1-\pi_i)\right)\left(\sum_{i=1}^{N} n_i \pi_i (1-\pi_i)\right) - \left(\sum_{i=1}^{N} x_i n_i \pi_i (1-\pi_i)\right)^2}\right].$$

Table 9.7 shows $\sqrt{Var(\hat{\pi})}$ for different x-values together with the estimated standard deviation obtained from 10 replications for logistic regression and 500 replications for sigmoid regression using the logistic regression function.

From this table we can see that the approximation of the standard deviations are almost identical to the estimated standard deviations obtained from the logistic regression. Sigmoid regression, however, gave two or three times as big standard deviations for all examined x-values compared to logistic regression.

9.2. Comparison between sigmoid regression and an ML estimation method proposed by Schmoyer.

To illustrate how Schmoyer's estimation method works we will use an example. We will also show the obtained results for this example using sigmoid regression. But first we will give a description of the estimation procedure.

9.2.1. The ML estimation method proposed by Schmoyer.

Suppose we have doses $x_1, \ldots, x_N$, administered to $n_1, \ldots, n_N$, subjects. Also suppose that there exists a sigmoid dose-response curve Π, such that $p_i = \Pi(x_i)$, i=1,...,N, where p_i denote the probability that an arbitrary subject responds in a specific way to a dose x_i. Let s_i, i=1,...,N, denote the slopes between two successive observations, that is,

$$s_i=\frac{p_i-p_{i-1}}{x_i-x_{i-1}}.$$

Denote the set of all **p** having non-negative unimodal slopes by **S**. The set S_i corresponds to sigmoid curves having modes between x_{i-1} and x_i. For each S_i a saddle point of the associated Lagrangian is determined. The log likelihood can be written as

$$\textbf{log lik}(\hat{\mathbf{p}}, \mathbf{p}) = \mathbf{c} + \sum_{i=1}^{N} (n_i\hat{p}_i\ln[p_i] + (1-\hat{p}_i)\ln[1-p_i]),$$

where $\hat{p}_i$ are the observed proportions and **c** is constant with respect to **p**. In the estimation each inequality constraint, $\mathbf{c}(\mathbf{p}) \geq 0$, is replaced with an equality constraint by introducing a slack variable s, such that $\mathbf{c}(\mathbf{p})-0.5s^2= 0$. Newton's method is used to find a zero of the gradient of the Lagrangian. This zero ensures a global maximum.

The computational algorithm is sensitive to starting values and Schmoyer recommends the following method to obtain adequate values:

Monotonize $\hat{\mathbf{p}}$ to $\tilde{\mathbf{p}}$. This is done using the pool-adjacent-violators algorithm with weights $w_i=n_i$ described in Barlow et al (1972). The monotonized values, $\tilde{p}_i$, are transformed to the corresponding slopes

$$\tilde{s}_i=\frac{\tilde{p}_i-\tilde{p}_{i-1}}{x_i-x_{i-1}}, \quad i=2,...,N.$$

For each $\tilde{S}_i$, i=2,...,N, the pool-adjacent-violators algorithm with weights $w_i=x_i-x_{i-1}$ is used to obtain a set of slopes consistent with S_i.

These slopes are then transformed back to proportions and used to start the optimization on S_i.
Unfortunately the algorithm fails to converge in some cases. But in these situations Schmoyer suggests an alternative estimate.

9.2.2. An example using Schmoyer's ML estimation method.
In his presentation of the estimation method Schmoyer demonstrates the procedure by giving the solution to the following experiment.
Rats were enclosed in chambers in which a specified dose of diesel fuel aerosol could be monitored and controlled. Let x_i be the doses of diesel fuel aerosol, p_i be the porpotion rats that died at the various doses and m_i be the number of rats tested for each x_i. The data from the experiment are shown in table 9.8.

TABLE 9.8.

x_i	8	16	24	28	32	48	64	72
m_i	30	40	40	10	30	20	10	10
$\hat{p}_i$	0.000	0.025	0.050	0.500	0.400	0.800	0.600	1.000

Table 9.8. The observations are obtained from an experiment, where x_i are the notation of the doses of diesel fuel aerosol, $\hat{p}_i$ the proportion rats that died at the various doses and m_i is the number of rats tested for each x_i.

From these observations we obtain the monotonized properties, $\tilde{p}_i$, (0.00, 0.025, 0.050, 0.425, 0.425, 0.733, 0.733, 1.000). The sigmoid ML estimates using the algorithm proposed by Schmoyer, $\tilde{p}_i(1)$ and our proposed estimation method, $\tilde{p}_i(2)$, with weights $\tilde{w}_i$ are shown in table 9.9.

TABLE 9.9.

x_i	8	16	24	28	32	48	64	72
m_i	30	40	40	10	30	20	10	10
$\tilde{p}_i(1)$	0.0000	0.0250	0.0500	0.3900	0.4480	0.6770	0.8921	0.9996
$\tilde{p}_i(2)$	0.0000	0.0250	0.0500	0.3792	0.4462	0.7138	0.7876	0.8244
$\tilde{w}_i$	0.0105	0.5743	0.2947	0.0156	0.0421	0.0329	0.0248	0.0051

Table 9.9. The estimates of the proportion dead rats are obtained using the Schmoyer estimation method and our proposed estimation method. Let x_i be the notation of the doses of diesel fuel aerosol and m_i the number of rats tested at different doses.

If we calculate the likelihood from Schmoyer's estimates we obtain $3.743*10^{-6}$. The corresponding value, calculated from the estimates obtained from our proposed estimation method, is $5.323*10^{-6}$. If we calculate the sum of square of errors instead we obtain for Schmoyer's estimation method SSE = 0.00353 and for our proposed estimation method SSE = 0.00159. To be able to calculate SSE for Schmoyer's method we have used sigmoid regression with $\tilde{\mathbf{p}}(1)$ as starting values in the iterations to obtain the corresponding weights, $\tilde{\mathbf{w}}(1)$ = (0.0104, 0.5694, 0.2922, 0.0146, 0.0421, 0.0317, 0.0361, 0.0035).

From ML and SSE we can see that our proposed method is superior to Schmoyer's at least in the above example. Another probable advantage of our proposed method is that in some cases Schmoyer's method does not converge. We haven't found any situation yet where our method does not converge.

9.2.3. General remarks.

From the example with lethality of diesel fuel aerosol smoke screens on rats we can see the difficulty to determine proper starting values in an iteration procedure. In his article Schmoyer proposes starting values which according to him are usually adequate. But in this case Schmoyer has found a local maximum. It is possible that his method in other situations will find the overall maximum.

When Schmoyer describes the method to obtain starting values in the iterations he uses the pool-adjacent-violators algorithm with weights $w_i=n_i$ described in Barlow et al (1972). This is an alternative estimation method to isotonic regression. It gives a good understanding of the estimation problem. In concave regression there is probably no analogous algorithm. This is due to the fact that concave regression uses fiction points in the estimation procedure. The following example illustrates the complexity of the concave estimation algorithm. Let two neighbouring bending points be denoted by $x_0^{(k_1)}$ and $x_0^{(k_2)}$ and the observations between them by x_i, $i=j+1,...,j+m_k$. Suppose we can prove that the observations between $x_0^{(k_1)}$ and $x_0^{(k_2)}$ have some property that made it impossible to obtain a bending point in the x-interval $[x_0^{(k_1)}, x_0^{(k_2)}]$ using only the corresponding y-values, $y_0^{(k_1)}$, $y_0^{(k_2)}$ and y_i, $i=j+1,.....,j+m_k$, with weights $w_0^{(k_1)}$, $w_0^{(k_2)}$ and w_i, $i=j+1,.....,j+m_k$.

Suppose we look at the overall problem with x-observations x_i, $i=1,.....,N$, and use the estimation procedure from both ends, x_1 and x_N, up to the bending points $x_0^{(k_1)}$ and $x_0^{(k_2)}$. The observations outside this x-interval will be represented by fiction points and there are no restrictions imposed on those by the behaviors in the interval. It is obvious that fiction points exist for which a concave function bending somewhere in the interval has smaller mean squares error than the LS straight line and thus there is a bending point in the inner of the interval for the overall problem.

10. COMPARISON BETWEEN THE PROPOSED ESTIMATION METHOD AND OTHER SIMILAR METHODS.

In this chapter we will examine properties of some other estimation methods and compare them to the method that we propose in this paper. We will mainly discuss bias, variances and the distributions of some estimates for monotonic increasing, unimodal and sigmoid regression functions.

10.1. Comparison between concave-up and isotonic regression.

In this chapter we will compare concave up regression with isotonic regression. Different properties will be examined for two different situations:

1) a monotonic increasing function that is concave with fix curvature, namely $y = -cx^2$ for $x \le 0$ and $c>0$

2) a monotonic increasing function $y = cx$ for $x \le 0$ and $c>0$.

The choice of regression functions depends on the fact that in the first situation concave-up regression will be in favour but in the second isotonic regression could be thought to fit very well. But first we start by giving a short description of the isotonic regression procedure.

10.1.1 The LSE-method of isotonic regression.

Isotonic regression is used for monotonic increasing or monotonic decreasing regression functions. A least squares estimate of the increasing regression function is obtained according to the following:

Suppose we have a probability distribution for each x_i, i=1,......,N. Also suppose that for each x_i a random sample $\{y_j(x_i), j=1,....,n(x_i)\}$ is available. The number of y-observations can be different for different x_i. The isotonic regression is identical to the regression function obtained from $\bar{y}(x_i)$ with weights

$$w(x_i) = \frac{n(x_i)}{\sum_{i=1}^{N} n(x_i)} .$$

Suppose that the x-observations are ordered, ie $x_1 < x_2 < .... < x_N$. If $\bar{y}(x_i)$ is monotone, then the averages themselves are the isotonic regression function. (The case when $\bar{y}(x_1) \geq \bar{y}(x_2) \geq ...... \geq \bar{y}(x_N)$ is treated analogously). If $\bar{y}(x)$ do not fulfill the monotonity restriction we compare the averages successively until the restriction is not fulfilled.

Suppose that the unequality $\bar{y}(x_k) \leq \bar{y}(x_{k+1})$ does not hold. We then estimate these two points according to

$$\bar{y}'(x_k) = \bar{y}'(x_{k+1}) = \frac{n(x_k)\bar{y}(x_k) + n(x_{k+1})\bar{y}(x_{k+1})}{n(x_k) + n(x_{k+1})} .$$

Before the procedure is continued we have to check if the monotonity restriction between $\bar{y}(x_{k-1})$, $\bar{y}'(x_k)$ and $\bar{y}'(x_{k+1})$ is fulfilled, ie

$$\bar{y}(x_1) \leq \; ..... \; \leq \bar{y}(x_{k-1}) \leq \bar{y}'(x_k) = \bar{y}'(x_{k+1}).$$

If this is not the case we have to re-estimate the y-value for x_{k-1}, x_k and x_{k+1};

$$\bar{y}''(x_{k-1}) = \bar{y}''(x_k) = \bar{y}''(x_{k+1}) =$$

$$\frac{n(x_{k-1})\bar{y}(x_{k-1}) + n(x_k)\bar{y}(x_k) + n(x_{k+1})\bar{y}(x_{k+1})}{n(x_{k-1}) + n(x_k) + n(x_{k+1})}.$$

Now we have to check if

$$\bar{y}(x_1) \leq \; ..... \; \leq \bar{y}(x_{k-2}) \leq \bar{y}''(x_{k-1}) = \bar{y}''(x_k) = \bar{y}''(x_{k+1}).$$

If this is not the case we have to re-estimate the y-values of x_{k-2}, x_{k-1}, x_k and x_{k+1} using the weighted average of the means. This is repeated until the monotonity restrictions are fulfilled for all the y-values corresponding to x_i, i=1,.....,k+1. We then continue to examine if the inequality $\bar{y}''(x_{k+1}) \leq \bar{y}(x_{k+2})$ holds and so on.

This procedure is repeated until estimates that fulfill the monotonity restrictions have been obtained for all x_i. Properties of isotonic regression are described in Barlow et al (1972) and Robertson et al (1989).

10.1.2. Comparison of some properties between concave-up and isotonic regression.

We will compare concave-up and isotonic regression both by some simulation studies and by theoretical results. But we start by examining some simulation results.

Suppose we have a regression function $\mu(x)$ with $y_i = \mu(x) + \varepsilon_i$, for each of N fixed equidistant x-values. Let $\mu(x_i)$ be a symmetric unimodal regression function with maximum in x = 0 and $\varepsilon_i \sim N(0, \sigma(x_i)^2)$, where $\sigma(x_i)^2$ are limited. The position of the nearest x-observation compared to x_{max} was randomized using a uniform distribution in the interval $[-0.5\Delta, 0.5\Delta]$ centered at x_{max}.

10.1.2.1. Comparision of regression functions with constant curvature. BIAS of $\hat{\mu}_i$.

To examine how the curvature of the regression function influences bias of the estimated μ_i, i=1,...,N, a simulation study was performed for the variation interval $0.25 \leq c \leq 3.0$. The regression functions examined were of the type $\mu(x) = -cx^2$. The corresponding x-interval was $-7 \leq x \leq 0$ and $\Delta=0.5$. The results from 25 000 replications concerning bias for the normalized curvature NC=0.25, 0.50 and 0.75 are shown in table 10.1.

The standard deviations of the means in the simulations for concave-up regression lie between 0.004 - 0.006 and for isotonic regression between 0.004 - 0.008 From this table we can see that the estimated bias obtained from both concave-up and isotonic regression vary slowly. We also get almost unbiased estimates of μ_i for all the examined cases. The largest difference is found for c=0.25 in the inner of the x-interval. For c=0.5 and c=0.75 the sizes of bias seem to be almost the same for both estimation methods except in one of the end points, x=0, where isotonic regression suffers from boundary affects.

TABLE 10.1 $\quad NC = \dfrac{c\Delta^2}{\sigma/\sqrt{n}}$

	Concave-up			Isotonic		
x_i	0.25	0.50	0.75	0.25	0.50	0.75
-7.0	-0.169	-0.108	-0.074	-0.021	0.021	0.024
-6.5	0.107	0.008	0.066	0.009	0.020	0.023
-6.0	0.063	0.028	0.014	0.016	0.028	0.031
-5.5	0.013	-0.003	-0.004	0.012	0.026	0.031
-5.0	-0.003	0.001	0.004	0.010	0.031	0.038
-4.5	-0.007	-0.002	-0.002	0.007	0.027	0.035
-4.0	-0.011	-0.011	-0.011	-0.012	0.006	0.016
-3.5	-0.011	-0.013	-0.014	-0.021	0.007	0.019
-3.0	-0.002	0.000	0.001	-0.038	-0.009	0.007
-2.5	-0.002	-0.002	-0.002	-0.052	-0.028	-0.008
-2.0	-0.006	-0.006	-0.008	-0.072	-0.051	-0.030
-1.5	-0.008	-0.006	-0.005	-0.075	-0.069	-0.053
-1.0	-0.022	-0.010	-0.008	-0.052	-0.091	-0.098
-0.5	-0.045	-0.054	-0.051	0.058	-0.016	-0.057
0.0	0.058	0.053	0.047	0.376	0.319	0.281

Table 10.1. Random numbers were generated so that $Y \sim N(-NCx_i^2;1)$ for 15 equidistant x-observations in the x-interval $-7 \leq x \leq 0$. Bias of $\hat{Y}_i$, i=1,..,15, were estimated for $0.25 \leq NC \leq 3.00$ using 25 000 replications in concave-up regression and isotonic regression.

VARIANCE of $\hat{Y}_i$.

Since the bias obtained from the two estimation methods differed so little, it was of course very interesting to examine the variance of $\hat{Y}_i$. This was done by simulations. In the simulation studies we examined the relation between $Var(Y_i)$ and σ^2. The results were rescaled and shown in table 10.2.

From this table we see that $Var(\hat{y}_i)/\sigma^2$ obtained from concave-up regression is bigger at the end points than in the middle of the examined x-interval. For the isotonic regression it is the other way round. But in the inner of the examined x-interval the size of the variance is almost twice as big for isotonic regression as for concave-up regression.

Another conclusion from the simulation results is that in the same way as with bias, $\mathrm{Var}(\hat{y}_i)/\sigma^2$ for both estimation methods seem to vary slowly.

TABLE 10.2. $NC = \dfrac{c\Delta^2}{\sigma/\sqrt{n}}$

x_i	Concave-up 0.25	Concave-up 0.50	Concave-up 0.75	Isotonic 0.25	Isotonic 0.50	Isotonic 0.75
-7.0	0.783	0.891	0.915	1.336	0.548	0.630
-6.5	0.489	0.589	0.684	1.140	0.731	0.512
-6.0	0.399	0.517	0.664	1.023	1.157	0.479
-5.5	0.375	0.523	0.626	0.912	1.562	0.634
-5.0	0.377	0.527	0.615	0.823	1.394	1.051
-4.5	0.366	0.498	0.617	0.748	1.224	1.601
-4.0	0.390	0.530	0.620	0.698	1.066	1.388
-3.5	0.401	0.528	0.609	0.604	0.939	1.207
-3.0	0.394	0.536	0.642	0.527	0.850	1.052
-2.5	0.406	0.532	0.611	0.459	0.734	0.922
-2.0	0.406	0.547	0.624	0.398	0.610	0.792
-1.5	0.404	0.518	0.639	0.348	0.499	0.631
-1.0	0.409	0.547	0.663	0.335	0.436	0.522
-0.5	0.413	0.509	0.580	0.368	0.437	0.488
0.0	0.751	0.729	0.795	0.528	0.587	0.627

Table 10.2. Random numbers were generated so that $Y \sim N(-NCx_i^2;1)$ for 15 equidistant x-observations in the x-interval $-7 \leq x \leq 0$. The variances of $\hat{Y}_i$, i=1,..,15, were estimated for $0.25 \leq NC \leq 3.00$ using 25 000 replications in concave-up regression and isotonic regression.

THE DISTRIBUTION of $\hat{Y}_i$.

From the simulations using 25 000 replications concerning bias and variance we also obtained the 3:rd, μ_3, and the 4:th, μ_4, moments for $\hat{Y}_i$ for concave-up and isotonic regression. The standardized skewness, μ_3/s^3, and the standardized kurtosis, μ_4/s^4, were determined. For NC=0.5 the standardized skewness was around -0.177 to 0.336 for concave-up regression and around -0.293-0.395 for isotonic regression. The standardized kurtosis were very near 3.0 for both the estimation methods except in the end point x=-7.0, where concave-up regression had $\mu_4/s^4 = 6.118$ and isotonic regression $\mu_4/s^4 = 9.631$. The tendency was the same for the other values of NC.

10.1.2.2. Comparision of symmetric regression functions with linear functions.

BIAS of $\hat{Y}_i$.

Another simulation study was performed to examine how bias was affected by the fact that the regression function consists of linear functions. The simulation study was performed so that we would be able to compare these results with the preceding study with constant curvature. The regression functions examined were of the type $\mu(x) = cx$ where the constant varied from 0.25 to 3.0. The simulation study was performed for 21 equidistant x-observations in the interval $-20 \le x \le 0$. The result from 25 000 replications for the normalized slope $N\beta = 0.25$, 0.50 and 0.75 is shown in table 10.3.

TABLE 10.3. $N\beta = \frac{c\Delta}{\sigma/\sqrt{n}}$

x_i	Concave-up 0.25	Concave-up 0.50	Concave-up 0.75	Isotonic 0.25	Isotonic 0.50	Isotonic 0.75
-20.0	-0.504	-0.505	-0.506	-0.299	-0.197	-0.133
-19.0	-0.159	-0.161	-0.162	-0.087	-0.027	-0.002
-18.0	-0.046	-0.048	-0.049	-0.027	0.003	0.009
-17.0	0.018	0.015	0.014	-0.004	0.009	0.012
-16.0	0.059	0.057	0.056	0.001	0.005	0.005
-15.0	0.089	0.088	0.087	0.005	0.008	0.008
-14.0	0.111	0.110	0.109	0.008	0.011	0.011
-13.0	0.126	0.126	0.125	0.008	0.012	0.015
-12.0	0.135	0.136	0.136	0.002	0.001	0.001
-11.0	0.140	0.141	0.141	0.004	0.003	0.003
-10.0	0.139	0.142	0.143	0.009	0.007	0.007
-9.0	0.134	0.138	0.140	0.016	0.005	0.004
-8.0	0.125	0.131	0.132	0.040	0.015	0.009
-7.0	0.111	0.118	0.121	0.104	0.051	0.030
-6.0	0.090	0.101	0.104	0.310	0.214	0.152
-5.0	0.062	0.076	0.081	1.190	2.227	3.334
-4.0	0.024	0.043	0.049	0.886	1.723	2.584
-3.0	-0.026	-0.005	0.004	0.625	1.223	1.834
-2.0	-0.095	-0.077	-0.065	0.370	0.723	1.084
-1.0	-0.186	-0.194	-0.196	0.117	0.223	0.334
0.0	-0.315	-0.397	-0.439	-0.137	-0.277	-0.416

Table 10.3. Random numbers were generated so that $Y \sim N(-N\beta x_i;1)$ for 21 equidistant x-observations in the x-interval $-20 \leq x \leq 0$. Bias of $\hat{Y}_i$, i=1,..,21, were estimated for $N\beta$ = 0.25, 0.50 and 0.75 using 25 000 replications in concave-up regression and isotonic regression.

In this simulation study we also examined how big the fractions between bias and the corresponding expected y-values were compared to the situation with constant curvature. In all cases with linear functions the fractions were bigger than for situations with constant curvature. The standard deviations of the means in the simulations for concave-up regression lie between 0.0016 - 0.0042 and for isotonic regression between 0.0006 - 0.0060.

From the table we see that we have the same tendency with estimates that vary slowly as for situations with constant curvature. The bias obtained from concave-up regression is bigger at the end points than in the middle of the examined x-interval. Bias obtained from isotonic regression does not behave in such a regular way. The size of bias seems to be a little bigger for concave-up regression except in the neighbourhood of x=0, where isotonic regression suffers from boundary affects. Bias for concave-up regression is bigger than bias for isotonic regression. The explaination is that concave regression always overestimates the observations in the middle of the examined x-interval and that the simulations are made for x-values on one side of a unimodal function.

VARIANCE of $\hat{Y}_i$.

The comparison of the two estimation methods concerning variance is also done by simulations. In the simulation studies we examined the relation between $Var(Y_i)$ and σ^2. The results were rescaled and shown in table 10.4. From this table we see that $Var(\hat{y}_i)/\sigma^2$ obtained from concave-up regression is slightly smaller than the corresponding variances obtained from isotonic regression. Another conclusion from the simulation results is that at the end point of the examined x-interval the variances seem to increase when the normalized slope increases. This tendency is very clear in isotonic regression for all x-values except near x=0. The obtained variances seem to vary slowly for both estimation methods.

From this table we can also see the same tendency for concave-up regression as for regression functions with constant curvature, ie the variances of $\hat{Y}_i$ compared to σ^2 are bigger at the end points of the regression curve. For isotonic regression there is no tendency like that.

TABLE 10.4. $N\beta = \frac{c\Delta}{\sigma/\sqrt{n}}$

x_i	Concave-up 0.25	Concave-up 0.50	Concave-up 0.75	Isotonic 0.25	Isotonic 0.50	Isotonic 0.75
-20.0	0.441	0.440	0.439	0.579	0.730	0.909
-19.0	0.211	0.211	0.210	0.387	0.518	0.702
-18.0	0.159	0.159	0.159	0.334	0.457	0.652
-17.0	0.131	0.130	0.130	0.325	0.457	0.636
-16.0	0.111	0.110	0.110	0.328	0.441	0.617
-15.0	0.097	0.096	0.096	0.323	0.432	0.601
-14.0	0.086	0.086	0.086	0.319	0.424	0.602
-13.0	0.079	0.078	0.078	0.311	0.430	0.580
-12.0	0.073	0.073	0.073	0.301	0.422	0.555
-11.0	0.070	0.070	0.070	0.299	0.412	0.553
-10.0	0.069	0.069	0.069	0.295	0.424	0.540
-9.0	0.070	0.070	0.070	0.301	0.428	0.533
-8.0	0.073	0.073	0.073	0.319	0.446	0.534
-7.0	0.077	0.079	0.079	0.373	0.479	0.575
-6.0	0.083	0.086	0.086	0.570	0.520	0.784
-5.0	0.092	0.096	0.097	0.072	0.709	0.008
-4.0	0.104	0.109	0.110	0.026	0.011	0.008
-3.0	0.121	0.125	0.127	0.019	0.009	0.008
-2.0	0.146	0.150	0.152	0.016	0.008	0.008
-1.0	0.181	0.197	0.201	0.014	0.008	0.008
0.0	0.226	0.286	0.327	0.012	0.008	0.008

Table 10.4. Random numbers were generated so that $Y \sim N(-N\beta x_i;1)$ for 21 equidistant x-observations in the x-interval $-20 \le x \le 0$. The variances of $\hat{Y}_i$, i=1,..,21, were estimated for $N\beta$ = 0.25, 0.50 and 0.75 using 25 000 replications in concave-up regression and isotonic regression.

THE DISTRIBUTION of $\hat{Y}_i$.

From the simulations using 25 000 replications concerning bias and variance we also obtained the 3:rd, μ_3, and the 4:th, μ_4, moments of $\hat{Y}_i$ for regression functions with linear functions. The standardized skewness, μ_3/s^3, and the standardized kurtosis, μ_4/s^4, were determined. For $N\beta=0.5$ the standardized skewness was around -0.605 to 0.309 for concave-up regression and around -0.365 to 2.105 for isotonic regression. The standardized kurtosis was very near 3.0. The tendency was the same for the other values of $N\beta$.

10.1.2.3. Comparison of theoretical results.

If we look at different consistency properties we can conclude that both estimation methods will converge to a least squares estimate within a finite number of steps. We have also proved consistency when the number of y-observations increases to infinity for each x_i. The set of isotonic functions is a convex cone. This is also true for concave-up functions.

10.1.2.4. Conclusions and remarks.

When we compare isotonic and concave-up regression we can conclude that none of the methods was much better than the other. If one method had a small bias in some situation it was often combined with a larger variance and so on. The standardized skewness and the standardized kurtosis for isotonic regression were also very close to the corresponding values of the concave-up regression. Therefore we conclude that these two estimation methods will be rather equal for regression functions with linear functions as well as for functions with constant curvature.

10.2. Comparison between concave and unimodal regression.

In this chapter we will compare concave regression to unimodal regression. Some theoretical results will be discussed. Different properties will be examined by simulations for two different situations:

1) a unimodal function that is concave, namely $y = -cx^2$

2) a unimodal function of the type

$$y = \begin{cases} cx & \text{for } x \leq 0 \\ -cx & \text{for } x > 0. \end{cases}$$

The regression functions are chosen for the same reasons as for monotonic increasing functions, namely that in the first situation concave regression will be in favour but in the second unimodal regression might be a better choice. But first we start by giving a short description of the unimodal regression procedure.

10.2.1. The LSE-method of unimodal regression.

If a partition in one non-descreasing and one non-increasing phase is known, then the least squares estimate of the function is obtained by using monotonic regression on each phase. But usually the partition is unknown. In Frisén (1980) a least squares estimation method is described for this situation according to the following:

First suppose that the regression function is not unimodal but monotonic decreasing. Thus we can imagine a partition to the left of the smallest x-observation. The least-squares-estimate for this case is obtained by using monotonic regression on all observations. The corresponding sum Q_1 of squares is calculated. Next suppose that the partition in a non-decreasing and a non-increasing phase is between x_1 and x_2. Calculate the monotonic regression estimate on each side of the partition. Calculate the corresponding sum Q_2 of squares. This procedure is repeated for all conceivable partitions and the sums Q_i, i=1,.....,N+1, of the squared deviations between observed and estimated values of $f(x_i)$ are calculated. The partition that gives the smallest value of Q_i, i=1,.....,N+1, will correspond to the least squares estimate $y^+(x_i)$ of $f(x_i)$. Suppose that the partition in monotonic phases, that gives the minimal sum of squares, is between x_{m-1} and x_m. The estimated maximum point the of unimodal regression function will be the largest of the estimates $y^+(x_{m-1})$ and $y^+(x_m)$. If for instance $y^+(x_m) > y^+(x_{m-1})$, then $\hat{x}_{max} = x_m$ and $\hat{y}_{max} = y^+(x_m)$.

More properties of unimodal regression are described in Dahlbom (1983,1986).

10.2.2. Comparison of some properties of concave and unimodal regression.

When we compare concave and unimodal regression we start by examining some simulation results.

Suppose we have a sample $y_{ij} = \mu(x_i) + \varepsilon_{ij}$, j=1,....,n($x_i$), for each of N fixed equidistant x-values, where $\mu(x_i)$ is a symmetric unimodal regression function with maximum in x = 0. Let $\varepsilon_i \sim N(0, \sigma(x_i)^2)$, where $\sigma(x_i)^2$ are limited. The position of the nearest x-observation compared to x_{max} was randomized using a uniform distribution in the interval [-0.5Δ, 0.5Δ] centered at x_{max}.

10.2.2.1. Comparison of estimates of regression functions with constant curvature.

BIAS of $Y(\hat{x}_{max})$

To examine how the curvature of the regression function influences bias of $Y(\hat{x}_{max})$ a simulation study was performed for the variation interval $0.20 \leq c \leq 2.50$. The regression functions examined were of the type $\mu(x) = -cx^2$. The corresponding x-interval was $-10 \leq x \leq 10$. The result from 25 000 replications concerning bias is shown in table 10.5.

TABLE 10.5.

$NC = \frac{c\Delta^2}{\sigma/\sqrt{n}}$	Concave $Y_{bias}(\hat{x}_{max})$	Unimodal $Y_{bias}(\hat{x}_{max})$
0.20	0.258	0.852
0.50	0.239	0.570
0.75	0.215	0.436
1.00	0.186	0.334
1.50	0.123	0.185
2.00	0.059	0.093
2.50	-0.013	0.041

Table 10.5. Random numbers were generated so that $Y \sim N(-NCx_i^2;1)$ for 21 equidistant x-observations in the x-interval $-10 \leq x \leq 10$. Bias of $\hat{Y}_{max}$ was estimated for $0.20 \leq NC \leq 2.50$ using 25.000 replications in concave regression and unimodal regression.

The standard deviations of the means in the simulations lie between 0.003 - 0.006 for the concave regression and between 0.043 - 1.251 for unimodal regression. From this table we see that bias obtained from unimodal regression is slightly bigger than bias obtained from concave regression. We can also see that the values of the normalized curvature in the interval $2.0 \le NC \le 2.50$ will give an almost unbiased estimate of y_{max} for both estimation methods. From the simulation results we can draw the conclusion that within the examined interval of NC bias obtained from both concave and unimodal regression decreases when c, Δ, and σ remain constant and the number of y-observations, n, for each x_i increases.

VARIANCE of $Y(\hat{x}_{max})$.

Another interesting property was the difference between the properties of the two estimation methods according to the variance of $\hat{Y}_{max}$. We started by examining the influence on variance by the

TABLE 10.6

	Concave	Unimodal
$NC = \frac{c\Delta^2}{\sigma/\sqrt{n}}$	$\frac{\hat{V}\hat{a}r(y_{max})}{\sigma^2}$	$\frac{\hat{V}\hat{a}r(y_{max})}{\sigma^2}$
0.20	0.352	0.464
0.50	0.485	0.555
0.75	0.549	0.600
1.00	0.602	0.640
1.50	0.683	0.731
2.00	0.746	0.827
2.50	0.793	0.906

Table 10.6. Random numbers were generated so that $Y \sim N(-NCx_i^2;1)$ for 21 equidistant x-observations in the x-interval $-10 \le x \le 10$. The variance of $\hat{Y}_{max}$ was estimated for $0.20 \le NC \le 2.50$ using 25 000 replications in concave regression and unimodal regression.

curvature. In the simulation studies we examined the relation between $Var(y_{max})$ and σ^2. The results were rescaled and shown in table 10.6.

From this table we see that $Var(\hat{Y}_{max})/\sigma^2$ obtained from unimodal regression is slightly bigger than the corresponding variance obtained from concave regression. Another conclusion from the simulation results is that when NC increases $Var(\hat{Y}_{max})/\sigma^2$ obtained from both estimation methods increase towards 1.

THE DISTRIBUTION of $Y(\hat{x}_{max})$.

From the simulations using 25 000 replications the 3:rd, μ_3, and the 4:th, μ_4, moments for both $\hat{Y}_{max}$ and $\hat{X}_{max}$ were estimated from concave and unimodal regression. The standardized skewness, μ_3/s^3, and the standardized kurtosis, μ_4/s^4, were determined. The results showed that for all values of NC the standardized skewness for $\hat{Y}_{max}$ was around 0.175 to 0.190 for concave regression and around 0.148 to 0.332 for unimodal regression. The standardized kurtosis was very near 3.0 for both the estimation methods.

BIAS of $\hat{X}_{max}$.

For both estimation methods we will obtain an unbiased estimate of x_{max}.

VARIANCE of $\hat{X}_{max}$.

In the simulation studies concerning the variance of the estimated x_{max} we examined the relation between $\hat{V}ar(\hat{x}_{max})$ and Δ^2. The results were rescaled and shown in table 10.7. From this table we can see that $\hat{V}ar(\hat{x}_{max})/\Delta^2$ is very much alike for both the estimation methods. For NC $\leq$ 1 it seems that the variance obtained from concave regression is slightly smaller. But for NC > 1

TABLE 10.7.

$NC = \frac{c\Delta^2}{\sigma/\sqrt{n}}$	Concave $\frac{\hat{V}ar(\hat{x}_{max})}{\Delta^2}$	Unimodal $\frac{\hat{V}ar(\hat{x}_{max})}{\Delta^2}$
0.20	1.411	1.440
0.50	0.634	0.646
0.75	0.443	0.459
1.00	0.347	0.368
1.50	0.244	0.237
2.00	0.190	0.139
2.50	0.159	0.071

Table 10.7. Random numbers were generated so that $Y \sim N(-NCx_i^2;1)$ for 21 equidistant x-observations in the x-interval $-10 \leq x \leq 10$. The variance of $\hat{X}_{max}$ was estimated for $0.20 \leq NC \leq 2.50$ using 25 000 replications in concave regression and unimodal regression.

it seems to be the other way round. We can also see that when NC increases $\hat{V}ar(\hat{x}_{max})/\Delta^2$ decreases for both estimation methods.

THE DISTRIBUTION of $\hat{X}_{max}$.

In the same simulation study as for $\hat{Y}_{max}$ we also determined the estimates of μ_3/σ^3 and μ_4/σ^4 for $\hat{X}_{max}$. Since the regression function was symmetric around $x=x_{max}$ we obtained $\mu_3/s^3 \approx 0.000$ for all values of NC. The simulation results also showed that for all values of NC the standardized kurtosis, μ_4/s^4, for $\hat{X}_{max}$ was around 2.500 to 2.660 for concave regression. For unimodal regression the obtained results are shown in table 10.8.

From this table we can see that μ_4/s^4 increases when NC increases.

TABLE 10.8.

$\frac{c\Delta^2}{\sigma/\sqrt{n}}$	0.20	0.50	0.75	1.00	1.50	2.00	2.50
$\frac{\mu_4}{s^4}$	2.612	2.612	2.450	2.548	4.202	7.194	14.081

Table 10.8. Random numbers were generated so that $Y \sim N(-NCx_i^2;1)$ for 21 equidistant x-observations in the x-interval $-10 \leq x \leq 10$. The standardized kurtosis of $\hat{X}_{max}$ was estimated for $0.20 \leq NC \leq 2.50$ using 25 000 replications in concave regression and unimodal regression.

10.2.2.2. Comparison of regression functions with piecewise linear functions.

BIAS of $Y(\hat{x}_{max})$

To be able to compare the results from this simulation study with the results from the preceding study using regression functions with constant curvature the examined situations were chosen as similar as possible. The variation interval for c was the same, namely $0.20 \leq c \leq 2.50$. The regression functions examined were of the type

$$\mu(x) = \begin{cases} cx & \text{for } x \leq 0 \\ -cx & \text{for } x > 0. \end{cases}$$

The corresponding x-interval was $-25 \leq x \leq 25$. The result from 25 000 replications concerning bias is shown in table 10.9. The standard deviations of the means in the simulations lie between 0.0024 - 0.0035 for the concave regression and between 0.0043 - 0.0060 for unimodal regression.

TABLE 10.9

$N\beta = \frac{c\Delta}{\sigma/\sqrt{n}}$	Concave $Y_{bias}(\hat{x}_{max})$	Unimodal $Y_{bias}(\hat{x}_{max})$
0.50	-0.467	0.705
0.75	-0.574	0.499
1.00	-0.667	0.358
1.50	-0.749	0.185
2.00	-0.897	0.091
2.50	-0.967	0.041

Table 10.9. Random numbers were generated so that $Y \sim N(-N\beta x_i;1)$ for 51 equidistant x-observations in the x-interval $-25 \leq x \leq 25$. Bias of $\hat{Y}_{max}$ was estimated for $0.50 \leq N\beta \leq 2.50$ using 25 000 replications in concave regression and unimodal regression.

From this table we see the same tendency as for regression functions with constant curvature. Bias obtained from unimodal regression is positive and bias obtained from concave regression is negative. We can see that bias is always bigger for regression functions with constant curvature compared to regression functions with piecewise linear functions. When we have regression functions with piecewise linear functions the estimate of y_{max} will be almost unbiased when $N\beta = 3.0$. The corresponding value for functions with constant curvature is $N\beta < 0.5$.

VARIANCE of $Y(\hat{x}_{max})$.

When we examined how the variance is influenced by the curvature the same simulation situations as for bias were used. The results from 25 000 replications were rescaled and shown in table 10.10.

TABLE 10.10.

$N\beta = \dfrac{c\Delta}{\sigma/\sqrt{n}}$	Concave $\dfrac{\hat{V}ar(\hat{y}_{max})}{\sigma^2}$	Unimodal $\dfrac{\hat{V}ar(\hat{y}_{max})}{\sigma^2}$
0.50	0.177	0.453
0.75	0.205	0.528
1.00	0.232	0.598
1.50	0.258	0.718
2.00	0.283	0.817
2.50	0.310	0.893

Table 10.10. Random numbers were generated so that $Y \sim N(-N\beta x_i;1)$ for 51 equidistant x-observations in the x-interval $-25 \leq x \leq 25$. The variance of $\hat{Y}_{max}$ was estimated for $0.50 \leq N\beta \leq 2.50$ using 25 000 replications in concave regression and unimodal regression.

From this table we can see that when $N\beta$ increases $Var(\hat{Y}_{max})/\sigma^2$ obtained from both estimation methods increases too.

THE DISTRIBUTION of $Y(\hat{x}_{max})$.

From the simulations using 25 000 replications the 3:rd, μ_3, and the 4:th, μ_4, moments for both $\hat{Y}_{max}$ and $\hat{X}_{max}$ were estimated from concave and unimodal regression. The standardized skewness, μ_3/s^3, and the standardized kurtosis, μ_4/s^4, were determined. The results showed that for all values of $N\beta$ the standardized skewness for $\hat{Y}_{max}$ was around -0.314 - -0.227 for concave regression and around 0.153 - 0.410 for unimodal regression. From this result we can see that the skewness differs more for unimodal regression than concave regression for different values of $N\beta$. The standardized kurtosis were around 2.90 - 3.20 for all values of Nβ for both the estimation methods.

<u>BIAS of $\hat{X}_{max}$.</u>

In this section we will obtain an unbiased estimate of x_{max} for both estimation methods.

<u>VARIANCE of $\hat{X}_{max}$.</u>

In the simulation studies concerning the variance of the estimated x_{max} we examined the relation between $\hat{V}ar(\hat{X}_{max})$ and Δ^2. The results were rescaled and shown in table 10.11.

TABLE 10.11.

$N\beta = \frac{c\Delta}{\sigma/\sqrt{n}}$	Concave $\frac{\hat{V}ar(\hat{X}_{max})}{\Delta^2}$	Unimodal $\frac{\hat{V}ar(\hat{X}_{max})}{\Delta^2}$
0.50	0.873	1.992
0.75	0.481	1.035
1.00	0.318	0.615
1.50	0.186	0.275
2.00	0.140	0.139
2.50	0.113	0.068

<u>Table 10.11.</u> Random numbers were generated so that $Y \sim N(-N\beta x_i;1)$ for 51 equidistant x-observations in the x-interval $-25 \leq x \leq 25$. The variance of $\hat{X}_{max}$ was estimated for $0.50 \leq N\beta \leq 2.50$ using 25 000 replications in concave regression and unimodal regression.

From this table we can see that when $N\beta \leq 2.00$ $\hat{V}ar(\hat{X}_{max})/\Delta^2$ is bigger for unimodal regression than for piecewise linear functions. For $N\beta > 2.00$ it is on the other way round. We can also see that when $N\beta$ increases $\hat{V}ar(\hat{X}_{max})/\Delta^2$ decreases for both estimation methods.

THE DISTRIBUTION of $\hat{X}_{max}$.

In the same simulation study as for $\hat{Y}_{max}$ we also determined the estimates of μ_3/σ^3 and μ_4/σ^4 for $\hat{X}_{max}$. Since the regression function was symmetric around $x=x_{max}$ we obtained $\mu_3/s^3 \approx 0.000$ for all values of $N\beta$. The simulations results also showed that for all values of $N\beta$ the standardized kurtosis for $\hat{X}_{max}$ was around 2.500 - 2.660 for both concave regression. For unimodal regression the obtained results are shown in table 10.12. From this table we can see that μ_4/s^4 increases when $N\beta$ increases.

TABLE 10.12.

$\frac{c\Delta}{\sigma/\sqrt{n}}$	0.50	0.75	1.00	1.50	2.00	2.50
$\frac{\mu_4}{s^4}$	4.011	4.193	4.577	5.739	8.281	15.030

Table 10.12. Random numbers were generated so that $Y \sim N(-N\beta x_i;1)$ for 51 equidistant x-observations in the x-interval $-25 \leq x \leq 25$. The standardized kurtosis of $\hat{X}_{max}$ was estimated for $0.50 \leq N\beta \leq 2.50$ using 25 000 replications in concave regression and unimodal regression.

10.2.2.3. Comparison of theoretical results.

If we look at different consistency properties we can conclude that we have an analogous situation compared to the preceding study concerning isotonic and concave-up regression functions. Thus both estimation methods will converge to a least squares estimate within

a finite number of steps. We have also proved consistency when the number of y-observations increases to infinity for each x_i.
The set of unimodal functions is a convex cone. This is also true for concave functions.

10.2.2.4. Conclusions and remarks.

When we compare unimodal and concave regression the following can be concluded:

Equalities: The standardized skewness and standardized kurtosis of $\hat{Y}_{max}$ are similar for unimodal and concave regression. Since all simulation studies contain a symmetric situation $\hat{X}_{max}$ is an unbiased estimate. The variances of $\hat{X}_{max}$ are bigger in concave regression for some values of NC and Nβ and bigger in unimodal regression for others. Since $\hat{X}_{max}$ is an unbiased estimate the standardized skewness of $\hat{X}_{max}$ = 0.000 for both unimodal and concave regression in all examined situations.

Differences: For big values of NC and Nβ bias of $\hat{Y}_{max}$ are comparable for both estimation methods in all the examined situations. But when NC or Nβ decreases to zero then, according to Dahlbom(1986), bias using unimodal regression increases to infinity while bias using concave regression has a limit value, c, where $0 \le c \le 0.150$. The variances of $\hat{Y}_{max}$, using concave regression, are smaller in all situations. The standardized kurtosis of $\hat{X}_{max}$ for concave regression are very close to the corresponding values of the Gaussian distribution for all values of NC and Nβ. This is not at all the case for unimodal regression. In this situation the standardized kurtosis are bigger than 3.0 for all values of NC and Nβ and show an increasing tendency when the normalized curvature or normalized slope increases. Therefore we conclude that the concave estimation method is superior compared to unimodal regression both for regression functions with piecewise linear functions as well as for functions with constant curvature.

10.3. Comparison between concave regression and an estimation method proposed by Fraser and Massam.

To illustrate how the Fraser and Massam estimation method works we will use an example. We will also show the obtained results for this example using concave regression. But first we will give a description of the estimation procedure.

10.3.1. The estimation method proposed by Fraser and Massam.

Suppose we have a regression function, $\eta(x_i)$, i=1,....,N. Then according to the concavity restriction the following holds

$$\frac{\eta(x_i)-\eta(x_{i-1})}{x_i-x_{i-1}} \geq \frac{\eta(x_{i+1})-\eta(x_i)}{x_{i+1}-x_i} \qquad i=1,.....,N-2.$$

This unequality can be rewritten according to

$$(x_{i+1}-x_i)\eta(x_{i-1}) + (x_{i-1}-x_{i+1})\eta(x_i) + (x_i-x_{i-1})\eta(x_{i+1}) \leq 0 \qquad (1)$$

Denote

$$\Omega = \{ \eta(x_i);\ \alpha_i'\eta(x_i) \leq 0,\ i=2,...,N-1\},$$

where $\alpha_i' = [0, 0, ..., (x_{i+1}-x_i), (x_{i-1}-x_{i+1}), (x_i-x_{i-1}), 0,...,0]$ with (i-1) zeros on the left and (N-i-2) on the right. We can also express (1) as $\gamma_i'\mathbf{M}\eta(x_i) \leq 0$, i=1,...,N-2, where $\mathbf{M}$ is the diagonal matrix (m_i) and

$$\gamma_i' = \mathbf{M}^{-1}\alpha_i', \qquad i=1,...,N-2 \qquad (2)$$

Thus $\mathbf{M}^{-1}$ = diag (m_i^{-1}). These new (N-2) vectors γ_i' will have the form

$$\gamma_i' = (0, 0, \ldots., (\frac{x_{i+1}-x_i}{m_i}), (\frac{x_{i-1}-x_{i+1}}{m_i}), (\frac{x_i-x_{i-1}}{m_i}), 0,\ldots,0).$$

Since we would like to use γ_i' to form a NxN matrix two new vectors, γ_{N-1}' and γ_N' are created according to

$$\gamma_{N-1}' = \frac{1}{\sqrt{\sum_{i=1}^{N} m_i}} (1, 1, 1,\ldots\ldots\ldots, 1)$$

$$\gamma_N' = \frac{1}{\sqrt{\sum_{i=1}^{N} m_i(x_i - \bar{x})^2}} (x_1 - \bar{x}, x_2 - \bar{x},\ldots\ldots\ldots, x_N - \bar{x})$$

where $\bar{x}$ is the weighted mean.

Denote the NxN matrix by **C**, ie **C** = $(\gamma_1, \gamma_2,\ldots\ldots, \gamma_N)$. Then according to (2) $\mathbf{C}'\mathbf{M}\eta$ will have its (N-2) first coordinates ≤ 0. Now let $\mathbf{B} = (\beta_1, \beta_2,\ldots\ldots, \beta_N)$ be a basis $\mathbf{C}'\mathbf{M}\mathbf{B} = -\mathbf{I}$ or $\mathbf{B} = -\mathbf{M}^{-1}(\mathbf{C}')^{-1}$; in particular $\beta_{N-1} = -\gamma_{N-1}$ and $\beta_N = -\gamma_N$ .Thus the points on the regression function can be expressed as

$$\eta = \sum_{i=1}^{N} b_i \beta_i$$

and the estimate, **y**, of η can be written as

$$\mathbf{y} = \sum_{i \in J} c_i \gamma_i + \sum_{i \in L-J} b_i \beta_i + b_{N-1}\beta_{N-1} + b_N \beta_N \quad \ldots\ldots\ldots\ldots(3)$$

with $c_i \geq 0$, $i \in J$ and $b_i > 0$, $i \in L$-J where $L = \{1,\ldots\ldots, N\text{-}2\}$ and J is a subset of L. The index values in the subset J are obtained according to the following:

First consider some initial point y_0 with the special restricted coordinates in (3) for some basis $\mathbf{B}_0$. Since an arbitrary point can be written as $y = \sum_{i=1}^{k} b_i\beta_i + \sum_{i=k+1}^{N} b_i\beta_i$ with $b_i > 0$, $i=1,\ldots\ldots,k$, we choose $y_0 = \sum_{i=1}^{k} b_i\beta_i$ which has the special restricted coordinates with respect to $\mathbf{B}_0 = \mathbf{B}$. If the data point y also has nonnegative coordinates with respect to B_0, ie if k=N, then $\mathbf{B}_0$ is the required basis and the point is its own estimate. If y has at least one negative coordinate with respect to $\mathbf{B}_0$, then we must continue the procedure. Denote b_i in the sum $y_0 = \sum_{i=1}^{k} b_i\beta_i$ by $b_i^{(1)}$.

Choose $b_j^{(0)} = \sum_{i=1}^{k} b_i$, $j=1,\ldots\ldots,N$. Denote $b_j^{(0)} = a_j^{(0)}$. For each $b_j < 0$

calculate

$$t_1 = \frac{a_j^{(0)}}{a_j^{(0)} - b_j^{(1)}} \qquad 0 \leq t_1 \leq 1.$$

The index value, j, that gives us the minimum t_1-value, indicates that the vector β_j should be excluded from the matrix $\mathbf{B}$ and replaced by γ_j. The new matrix is denoted $\mathbf{B}^1$. Now we consider

$$y = c_j^{(2)}\gamma_j + \sum_{i \neq j} b_i^{(2)}\beta_i.$$

If $c_j^{(2)} > 0$, and $b_i^{(2)} > 0$, we have found the estimate of the regression function. If we obtain some $b_i^{(2)} < 0$, $i=1,\ldots\ldots,N\text{-}2$, we must calculate

$$t_2 = \min \frac{a_i^{(1)}}{a_i^{(1)} - b_i^{(2)}} \qquad 0 \le t_2 \le 1,$$

where

$$a_i^{(1)} = t_1 b_i^{(1)} + (1-t_1) a_i^{(0)}.$$

Notice that the restriction of b_i is not examined for b_{N-1} and b_N. The index, k, that gives us the minimum t_2-value indicates that β_k in the matrix $\mathbf{B}^1$ should be replaced by γ_k. The procedure successively replaces β_i-values by γ_i if the corresponding b_i, $b_i < 0$, gives us the minimum t-value. The procedure also replaces γ_i-values by β_i if the corresponding c_i, $c_i < 0$, gives us the minimum t-value.

Some properties of the estimation method are described in Fraser and Massam (1987,1989).

10.3.2. An example using The Fraser and Massam estimation method.

In their presentation of the estimation method Fraser and Massam (1987) demonstrate the procedure by giving the solution of an example.

Let y_i be the average corn yield, x_i be the amount of nitrogen fertilizer and m_i be the number of y-observations for each x_i. The data are

x_i	0	20	40	60	80	120	160	180
m_i	27	9	8	10	9	19	10	8
y_i	22.94	41.58	65.46	58.81	81.74	82.15	96.59	94.01

First we can conclude that the matrix $\mathbf{M}$ will be

$$\begin{matrix} 27 & 0 & 0 & 0 & 0 & 0 & 0 & 0 \\ 0 & 9 & 0 & 0 & 0 & 0 & 0 & 0 \\ 0 & 0 & 8 & 0 & 0 & 0 & 0 & 0 \\ 0 & 0 & 0 & 10 & 0 & 0 & 0 & 0 \\ 0 & 0 & 0 & 0 & 9 & 0 & 0 & 0 \\ 0 & 0 & 0 & 0 & 0 & 19 & 0 & 0 \\ 0 & 0 & 0 & 0 & 0 & 0 & 10 & 0 \\ 0 & 0 & 0 & 0 & 0 & 0 & 0 & 8 \end{matrix}$$

The α-vectors, which are formed according to

$$\alpha_i' = (0, 0, ..., (x_{i+1}-x_i), (x_{i-1}-x_{i+1}), (x_i-x_{i-1}), 0,...,0),$$

will lead to the following values

$$\begin{aligned} \alpha_1' &= (20, -40, 20, 0, 0, 0, 0, 0) \\ \alpha_2' &= (0, 20, -40, 20, 0, 0, 0, 0) \\ \alpha_3' &= (0, 0, 20, -40, 20, 0, 0, 0) \\ \alpha_4' &= (0, 0, 0, 40, -60, 20, 0, 0) \\ \alpha_5' &= (0, 0, 0, 0, 40, -80, 40, 0) \\ \alpha_6' &= (0, 0, 0, 0, 0, 20, -60, 40). \end{aligned}$$

Using these results for $\gamma_i' = \mathbf{M}^{-1}\alpha_i'$ we will obtain

$$\gamma_1' = (\frac{20}{27}, \frac{-40}{9}, \frac{20}{8}, 0, 0, 0, 0, 0)$$

$$\gamma_2' = (0, \frac{20}{9}, \frac{-40}{8}, \frac{20}{10}, 0, 0, 0, 0)$$

$$\gamma_3' = (0, 0, \frac{20}{8}, \frac{-40}{10}, \frac{20}{9}, 0, 0, 0)$$

$$\gamma_4' = (0, 0, 0, \frac{40}{10}, \frac{-60}{9}, \frac{20}{19}, 0, 0)$$

$$\gamma_5' = (0, 0, 0, 0, \frac{40}{9}, \frac{-80}{19}, \frac{40}{10}, 0)$$

$$\gamma_6' = (0, 0, 0, 0, 0, \frac{20}{19}, \frac{-60}{10}, \frac{40}{8}).$$

The vectors γ_{N-1}' and γ_N' are created according to

$$\gamma_7' = \frac{1}{\sqrt{100}} (1, 1, 1, 1, 1, 1, 1, 1) = (0.1, 0.1, 0.1, 0.1, 0.1, 0.1, 0.1, 0.1)$$

$$\gamma_8' = \frac{1}{\sqrt{387348}} (-71.4, -51.4, -31.4, -11.4, 8.6, 48.6, 88.6, 108.6) = (-0.1147222,$$

-0.082587, -0.050452, -0.018317, 0.0138108, 0.078088, 0.142358, 0.1744934).

The γ-vectors will form the matrix **C**. By inversion of **C**$'$, multiplication of $\mathbf{M}^{-1}$, we obtain the matrix **B**:

-0.0188081	-0.0288709	-0.0326283	-0.0151698	-0.0045455	-0.0019866	-0.10	0.1142341
0.0262362	0.0100286	-0.0005195	-0.0026572	-0.0017316	-0.0008699	-0.10	0.0822358
0.0212805	0.0489280	0.0315893	0.0098553	0.0010822	0.0002468	-0.10	0.0502374
0.0163248	0.0378273	0.0636980	0.0223679	0.0038961	0.0013635	-0.10	0.0182391
0.0113690	0.0267267	0.0458070	0.0348804	0.0067100	0.0024802	-0.10	-0.0138108
0.0014576	0.0045255	0.0100240	0.0099055	0.0123377	0.0047136	-0.10	-0.0780880
-0.0084538	-0.0176757	-0.0257582	-0.0150694	-0.0070346	0.0069470	-0.10	-0.1423580
-0.0134095	-0.0287763	-0.0436490	-0.0275569	-0.0167208	-0.0169363	-0.10	-0.1744930

Since each data point on the regression function can be expressed as $\eta = \sum_{i=1}^{8} b_i\beta_i$ this matrix is, for each $b_i < 0$, i=1,......,6, used to calculate

$$t_1(i) = \frac{a_i^{(0)}}{a_i^{(0)} - b_i^{(1)}} \qquad 0 \le t_1 \le 1.$$

The index value, j, that gives us the minimum t_1-value, indicates that the vector β_j should be excluded from the matrix **B** and replaced by γ_j. Multiple linear regression gives us the following $b^{(1)}$-values

(-104.801375, 610.600136, -591.233584, 909.636852,. -561.089913, 391.991632,

-611.492827, -261.003189)

As $a_i^{(0)}$ we can choose $\sum_{i \in J} b_i = 1912.2286$, where $L = \{2, 4, 6\}$.Then, for each $b_i < 0$, we can calculate a t-value:

$$t_1(1) = 0.948 \qquad t_1(3) = 0.764 \qquad t_1(5) = 0.773.$$

This calculation gives us min $t_1(3)=0.764$ and the index-value=3. The 3:rd column in the matrix **B** is replaced by γ_3. Denote the new matrix by $\mathbf{B}^1$. Multiple linear regression in this situation gives us a new vector with $b^{(2)}$-values.

(11.388211, 192322677, 2.323756, 228.100438, -354.590860, 391.999681,

-611.494443, -261.003789)

In this vector b_5=-354.590860 is the only b_i-value, i=1,..,6, < 0. Therefore we replace the 5:th column in the matrix $\mathbf{B}^1$ by γ_5. Denote this new matrix by $\mathbf{B}^2$. We also calculate the corresponding t-value using the connection

$$a_i^{(1)} = 0.764\ b_i^{(1)} + 0.236\ a_i^{(0)}\ .$$

This gives us the t-value 0.060. Multiple linear regression in this situation gives us a new vector with b-values.

(1.773262, 226.935050, 2.131468, 91.312793, 0.551141, 213.314028,

- 611.497377, -261.004379)

This vector contains no b_i-value, i=1,..,6, < 0. The solution gives us the estimated regression function according to

y_i	$\hat{y}_i$	w_i
22.94	22.8680	27
41.58	41.5779	9
65.64	60.1993	8
58.81	67.4738	10
81.74	74.6552	9
82.15	84.4988	19
96.59	94.3267	10
94.01	93.9078	8

The sum of squares is 1580.030.

According to some unfortunate reason Fraser and Massam (1987) have obtained another result in their chapter. The matrix **B**, $\mathbf{B}^*$, that they use in their example is

1	2	3	4	5	6	7	8
-0.293	-0.487	-0.520	-0.366	-0.118	-0.479	-0.100	0.114
0.409	0.169	-0.008	-0.064	-0.045	-0.021	-0.100	0.082
0.331	0.828	0.504	0.238	0.028	0.006	-0.100	0.050
0.254	0.638	1.016	0.539	0.101	0.033	-0.100	0.018
0.177	0.451	0.731	0.841	0.174	0.060	-0.100	-0.014
0.023	0.763	0.160	0.239	0.320	0.114	-0.100	-0.078
-0.132	-0.298	-0.411	-0.363	-0.183	0.168	-0.100	-0.142
-0.209	-0.485	-0.696	-0.664	-0.434	-0.408	-0.100	-0.174

That this is not the right matrix is seen by

$$\mathbf{C'MB}^* = \begin{matrix} -15.600 & 0.060 & 0.000 & 0.000 & 0.000 & -8.620 & 0.000 & 0.000 \\ 0.020 & -16.980 & 0.000 & -0.020 & 0.000 & 0.000 & 0.000 & 0.000 \\ 0.000 & 0.000 & 0.060 & -15.940 & 0.020 & 0.000 & 0.000 & 0.000 \\ 0.000 & 0.000 & 13.720 & -0.020 & -24.120 & 0.000 & 0.000 & 0.000 \\ 0.000 & -0.040 & -54.920 & 11.880 & 0.000 & -25.960 & 0.000 & 0.000 \\ 0.020 & 13.740 & -17.800 & 0.000 & 0.020 & -24.120 & 0.000 & -0.020 \\ 0.000 & 1.307 & 0.299 & 0.000 & 0.000 & -1.162 & -1.000 & 0.000 \\ 0.000 & 1.019 & 0.422 & 0.000 & 0.000 & 1.337 & 0.000 & -1.001 \end{matrix} \neq -\mathbf{I}.$$

According to their calculations the use of the matrix $\mathbf{B}^*$ leads to the estimate, $\hat{y}$, of $\eta(x_i)$

$$\hat{y} = (22.98,\ 41.4984,\ 60.1084,\ 67.3106,\ 74.5317,\ 84.4668,\ 94.402,\ 94.0346)$$

with the weighted sum of squares 1569.33. Unfortunately this result is not concave since when we calculate

$$\lambda_i = \frac{\eta(x_i)-\eta(x_{i-1})}{x_i - x_{i-1}} \qquad i=1,\ldots\ldots,N-1$$

we obtain

$$\lambda_1 = \frac{41.4984 - 22.98}{20} = 0.92592$$

$$\lambda_2 = \frac{60.1084 - 41.4984}{40 - 20} = 0.9305$$

$$\lambda_3 = \frac{67.3106 - 60.1084}{60 - 40} = 0.36011$$

$$\lambda_4 = \frac{74.5317 - 67.3106}{80 - 60} = 0.361055$$

$$\lambda_5 = \frac{84.4668 - 74.5317}{120 - 80} = 0.248378$$

$$\lambda_6 = \frac{94.402 - 84.4668}{160 - 120} = 0.248380$$

$$\lambda_7 = \frac{94.0346 - 94.402}{180 - 160} = -0.018370.$$

This solution does not fulfill the concavity restriction.

Generally we can conclude that theoretically the Fraser and Massam estimation method probably leads to the least squares estimate but the inversion of the matrix **C**, together with the solving of the big equation system, estimating b_i, leads to numerical instability in the results. This is clearly seen in the data point $y(x_8) = 94.01$, which, according to the concavity and the least squares restrictions, should be estimated by itself, ie $\hat{y}(x_8) = 94.01$. Fraser and Massam's own evaluation of the procedure gives the estimate $\hat{y}(x_8) = 94.0346$ and when we used Fraser and Massam's procedure we obtained $\hat{y}(x_8) = 93.9071$.

These arguments lead to the conclusion that the estimation method proposed by Fraser and Massam is numerically instable. In practice it does not give the least squares estimate of the regression function. Sometimes they even obtain an estimate that does not fulfill the concavity restrictions.

10.3.3. The solution obtained from our proposed estimation method.

First we estimate an ordinary least squares regression line from the observations $\hat{y}(x) = 34.43 + 0.3310x$. Now, for each x_i we calculate a ρ-value. If any ρ-value > 0 this indicates that we have not found the least squares estimate.

$$\rho(x) = \left(0.00,\ 59.69,\ 56.91,\ 45.96,\ 38.05,\ 19.87,\ 5.798,\ 0.00\right)$$

The largest ρ-value is $\rho(x_2) = 59.69$. This indicates that we should include $x_2=20$ as a bending point. Now we have to estimate two straight lines connected in x_2. From the linear regression through a predetermined point we obtain

$$\hat{y}(x) = 22.94 + 1.5490x \qquad 0 \le x \le 20$$

$$\hat{y}(x) = 48.92 + 0.2505x \qquad 20 \le x \le 180.$$

When we compare the regression coefficients we can see that the concavity restriction is fulfilled. Therefore we continue with the inclusion part to examine if any other point should be included in the regression procedure as a bending point. For each x_i, we calculate a ρ-value.

$$\rho(x) = (0.00,\ 0.00,\ 10.27,\ 10.04,\ 11.90,\ 7.111,\ 2.604,\ 0.000)$$

The largest ρ-value is $\rho(x_5)= 11.90$, ie $x_5 = 80$ is included as a bending point. Now we have to estimate three straight lines connected in x_2 and in x_5. From the linear regression procedure we obtain

$$\hat{y}(x) = 22.94 + 1.0810x \qquad 0 \leq x \leq 20$$

$$\hat{y}(x) = 33.53 + 0.5513x \qquad 20 \leq x \leq 80$$

$$\hat{y}(x) = 64.52 + 0.1638x \qquad 80 \leq x \leq 180.$$

When we compare the regression coefficients we can see that the concavity restriction is fulfilled. Therefore we continue with the inclusion part to examine if a third point should be included in the regression procedure as a bending point. For each x_i, we calculate a ρ-value.

$$\rho(x) = (0.00,\ 0.00,\ 2.472,\ -1.510,\ 0.00,\ -0.600,\ 0.581,\ 0.00)$$

The largest ρ-value is $\rho(x_3)= 2.472$, ie $x_3 = 40$ is the new bending point. From the linear regression procedure with bending points in x_2, x_3 and x_5 we obtain

$$\hat{y}(x) = 22.94 + 0.9320x \qquad 0 \le x \le 20$$

$$\hat{y}(x) = 23.46 + 0.9059x \qquad 20 \le x \le 40$$

$$\hat{y}(x) = 43.03 + 0.4167x \qquad 40 \le x \le 80$$

$$\hat{y}(x) = 62.25 + 0.1765x \qquad 80 \le x \le 180.$$

From this result we can see that the regression coefficients fulfill the concavity restriction. We continue the inclusion part. For each x_i, we calculate a ρ-value.

$$\rho(x) = (0.00,\ 0.00,\ 0.00,\ -2.721,\ 0.00,\ -0.120,\ 0.725,\ 0.00)$$

The only ρ-value > 0 is $\rho(x_7)$= 0.7247 and x_7 = 160 is included as a bending point. From the linear regression procedure we obtain

$$\hat{y}(x) = 22.94 + 0.9320x \qquad 0 \le x \le 20$$

$$\hat{y}(x) = 23.83 + 0.9274x \qquad 20 \le x \le 40$$

$$\hat{y}(x) = 45.70 + 0.3606x \qquad 40 \le x \le 80$$

$$\hat{y}(x) = 54.72 + 0.2479x \qquad 80 \le x \le 160$$

$$\hat{y}(x) = 97.40 - 0.01882x \qquad 160 \le x \le 180.$$

If we continue the inclusion part and calculate a ρ-value for each x_i then we can se that no obtained ρ-value > 0 and all b-values fulfill the concavity restriction. Thus we have found the least squares estimate which gives ut the estimate of the regression function

$$\hat{y}(x) = (22.94,\ 41.58,\ 60.13,\ 67.13,\ 74.55,\ 84.47,\ 94.39,\ 94.01).$$

The least squares estimate gives us the sum of squares 1570.812. We can also see that the data point $y(x_8) = 94.01$, is estimated by itself, ie $\hat{y}(x_8) = 94.01$. As we concluded in the discussion concerning the Fraser and Massam procedure this is necessary if the least squares restrictions should hold.

10.4. Comparison between concave regression and an estimation method proposed by Wu.

To be able to compare the method by Wu with our proposed method for concave regression under the same conditions as for Fraser and Massam's estimation method we will use the same disposition as in chapter 10.3. Thus we will start by giving a short description of the estimation method. Then we will discuss the solution to the same example as in the preceding chapter.
The least squares estimation method proposed by Wu involves a quadratic programming problem which makes it difficult to implement. Therefore he also gives an approximate solution to the estimation problem which is easy to use. This makes it very interesting and some properties of this approximation are examined.

10.4.1. The estimation method proposed by Wu.

Suppose we have m_i independent measures y_{ij}, taken at x_i, with the mean $\mathbf{u}(x_i)$ and the common variance σ^2, where $x_1 <....< x_N$. The problem is to estimate $\mathbf{u}(x_i)$ subject to the concavity constraints

$$\frac{\mathbf{u}(x_i)-\mathbf{u}(x_{i-1})}{x_i-x_{i-1}} \geq \frac{\mathbf{u}(x_{i+1})-\mathbf{u}(x_i)}{x_{i+1}-x_i} \qquad i=2,.....,N-1.$$

The least squares estimate of **u** under the concavity assumptions is obtained by solving

$$\min \sum_{i=1}^{N} m_i(u(x_i) - \bar{y}(x_i))^2$$

subject to the mentioned concavity constraints. According to the Kuhn-Tucker-Lagrange conditions, described e.g in Winston (1991), the solution $\hat{\mathbf{u}}$ is characterized by the following equations:

1) $\hat{U}_i' = Y_i' - \hat{\lambda}_i, \ \hat{\lambda}_i \geq 0, \ i=1,....,N-2.$

2) $\hat{U}_i' = Y_i',$ $\qquad i=N-1, N.$

3) $\hat{\lambda}_i = 0$ if $\dfrac{u(x_{i+1})-u(x_i)}{x_{i+1}-x_i} \geq \dfrac{u(x_{i+2})-u(x_{i+1})}{x_{i+2}-x_{i+1}},$

4) $\dfrac{u(x_2)-u(x_1)}{x_2-x_1} \geq ..... \geq \dfrac{u(x_N)-u(x_{N-1})}{x_N-x_{N-1}},$

where

$$\hat{U}_i' = \sum_{j=1}^{i}(x_{i+1} - x_j)m_j\hat{u}_j = \sum_{r=1}^{i}\Delta x_r U_r',$$

$$Y_i' = \sum_{j=1}^{i}(x_{i+1} - x_j)m_j\bar{y}_j = \sum_{r=1}^{i}\Delta x_r Y_r',$$

$$U_r' = \sum_{j=1}^{r} m_j\hat{u}_j, \quad Y_r' = \sum_{j=1}^{r} m_j\bar{y}_j; \quad \Delta x_r = x_{r+1} - x_r$$

and $\hat{\lambda}_i$ is uniquely determined by 1) - 4). Before giving the algorithm for concave regression a few terms must be defined:

The concavity constraint on (i-1, i, i+1) is called

<u>inactive</u> if $\frac{u(x_i)-u(x_{i-1})}{x_i-x_{i-1}} > \frac{u(x_{i+1})-u(x_i)}{x_{i+1}-x_i}$,

ie if the condition is strictly concave,

<u>active</u> if $\frac{u(x_i)-u(x_{i-1})}{x_i-x_{i-1}} = \frac{u(x_{i+1})-u(x_i)}{x_{i+1}-x_i}$,

ie if the condition is linear,

<u>violated</u> if $\frac{u(x_i)-u(x_{i-1})}{x_i-x_{i-1}} < \frac{u(x_{i+1})-u(x_i)}{x_{i+1}-x_i}$,

ie if the condition is strictly convex.

If the constraint on (i-1, i, i+1) is violated, then we can define a unique block B containing (i-1, i, i+1) such that:

a) any two consecutive points in B are related to each other by at least one violated or active constraint, and

b) the two extreme boundary points in B are related to their neighbouring points outside B only by inactive constraints.

Now we can give the algorithm of concave regression:

Step 0: $\hat{u}_i = \bar{y}_i$, i=1,...,N.

Step 1: Check if there is any violated constraint. If none, stop; otherwise denote the points in a violated constraint by i_0-1, i_0, i_0+1, go to step 2.

Step 2: Let B be the block of points containing i_0-1, i_0, i_0+1, as defined earlier. Replace $\{ \hat{u}_i \}_{i \in B}$ by the solution $\{ \hat{\mu}_i \}_{i \in B}$ of the constrained minimization problem:

$$\min_{\mu} \sum_{i \in B} m_i (\mu_i - \hat{u}_i)^2, \; \{ \mu_i \}_{i \in B}$$

subject to the concavity constraints on μ. Update $\{ \hat{u}_i \}$, $i=1,....,N$ and the status of each constraint.

Go to step 1.

Wu conjectures that the algorithm terminates in a finite number of steps. But he is not able to prove it. The convergence to the least squares solution can only be proved if

1) the algorithm eliminates all the violated constraints in a finite number of steps, and

2) if a violated or active constraint becomes active after one iteration, it should not be inactive at the end.

Thus when the estimation method converges it has the same properties as our proposed estimation method.

In the example discussed in the section concerning Fraser and Massam's estimation method, Wu's algorithm converges to the least squares solution. However Wu's estimation method involves a quadratic programming problem on a subset of variables and may not be easy to implement.

More properties of the estimation method proposed by Wu are described in Wu (1982).

10.4.2. The approximation proposed by Wu.

The least squares solution described in the preceding section has no closed form, because some of the constraints involving three consecutive points in B may be inactive. But when all the constraints involved in B are either violated or active the solution is obtained by fitting a weighted least squares line to $\{ \hat{u}_i \}_{i\in B}$. This is a fact that Wu uses in his alternative algorithm. It is very easy to use and therefore it is of great interest to examine the approximate method.

Denote the N-2 constraints by $C_1,\ldots\ldots,C_{N-2}$. If a constraint C_i is violated, there is a largest block of constraints $C_{i-f},\ldots\ldots\ldots,C_{i+g}$ containing C_i, such that all C_j, $i-f \le j \le i+g$, are either violated or active. The block C of points involved in C_j, $i-f \le j \le i+g$, is contained in the block B of points defined in Wu's least squares estimation method. Using these notations we can give the approximate algorithm according to the following:

Step 0: $\hat{u}_i = \bar{y}_i$, $i=1,\ldots,N$.

Step 1: Check if there is any violated constraint. If none, stop; otherwise denote the points in a violated constraint by i_0-1, i_0, i_0+1, go to step 2.

Step 2: Let C be the block of points defined above. Replace $\{ \hat{u}_i \}_{i\in C}$ by the solution $\{ \hat{\mu}_i \}_{i\in C}$, where $\hat{\mu}_i = \bar{u} + \beta(x_i - \bar{x})$,

$$\bar{u} = \frac{\sum_{i\in C} m_i \hat{u}_i}{\sum_{i\in C} m_i}, \quad \bar{x} = \frac{\sum_{i\in C} m_i x_i}{\sum_{i\in C} m_i} \quad \text{and} \quad \hat{\beta} = \frac{\sum_{i\in C} m_i(\hat{u}_i - \bar{u}_i)(x_i - \bar{x}_i)}{\sum_{i\in C} m_i(x_i - \bar{x}_i)^2}.$$

Update { $\hat{u}_i$ }, i=1,....,N and the status of each constraint. Go to step 1.

This estimation method terminates within N-2 steps.

Suppose that, in each iteration, no constraint involving any three consecutive points in B of step 2 is inactive. Then it is obvious that the algorithm is identical to Wu's least squares algorithm described in the preceding section.
Wu gives no other properties of the approximation method.

10.4.2.1. An example using Wu's approximation method.
To illustrate how the approximation works we use the same example as in the section describing Fraser and Massam's estimation method:
Let y_i be the average corn yield, x_i be the amount of nitrogen fertilizer and m_i be the number of y-observations for each x_i. The data are

x_i	0	20	40	60	80	120	160	180
m_i	27	9	8	10	9	19	10	8
$\bar{y}_i$	22.94	41.58	65.46	58.81	81.74	82.15	96.59	94.01
$\frac{\bar{y}_{i+1}-\bar{y}_i}{x_{i+1}-x_i}$	0.93	1.19	-0.33	1.15	0.01	0.36	-0.13	

From this table we see that the first violated constraint which should be selected is x_i = (40,60,80). In these three points the $\bar{y}_i$-values are replaced by the weighted least squares fitted values 59.65, 68.11 and 76.57. These values are the { $\hat{u}_i$ }, i=3,.....,5, in the first iteration. We then calculate $(\hat{u}_{i+1}-\hat{u}_i)/(x_{i+1}-x_i)$ and compare the constraints.

x_i	0	20	40	60	80	120	160	180
$\hat{u}_i^{(1)}$	22.94	41.58	59.65	68.11	76.57	82.15	96.59	94.01
$\frac{\hat{u}_{i+1}-\hat{u}_i}{x_{i+1}-x_i}$	0.93	0.90	0.42	0.42	0.35	0.36	-0.13	

Now the only constraint that is violated is x_i = (80,120,160). Again these three $\bar{y}_i$- values are replaced by the weighted least squares fitted values. The new values are 74.24, 84.36 and 94.49. These values are the $\{\hat{u}_i\}$, i=5,6,7, in the second iteration. This gives the following result:

x_i	0	20	40	60	80	120	160	180
$\hat{u}_i^{(2)}$	22.94	41.58	59.65	68.11	74.24	84.36	94.49	94.01
$\frac{\hat{u}_{i+1}-\hat{u}_i}{x_{i+1}-x_i}$	0.93	0.90	0.42	0.31	0.25	0.25	-0.02	

From this table we see that we have obtained the desired result. The sum of squares is 1778.1, which is a little higher than the sum of squares of the least squares solution, 1570.8. Its low sum of squares together with the fact that it is very easy to use, makes the approximation method a very practical and useful tool in concave regression.

10.4.2.2. Some properties of Wu's approximation method.

It is obvious that it is very important in which order the violated constraints are chosen. This is illustrated by the following example:

<u>Alternative 1:</u> Let {I} denote the indexes of the violated constraints. The first violated constraint C_i is chosen by finding the largest $b_{i+1} - b_i$, where i, $i+1 \in \{I\}$. This violated constraint will help us to find C, the <u>largest</u> block of constraints $C_{i-f}, \ldots\ldots, C_{i+g}$ containing C_i, such that all C_j, $i-f \leq j \leq i+g$, are either violated or active. Then in the corresponding x-points $(x_{i-f}, \ldots\ldots x_{i+g})$ the $\bar{y}_i$-values are replaced by the weighted least squares fitted values. These values are the $\{\hat{u}_i\}$, in the first iteration. We then calculate $(\hat{u}_{i+1}-\hat{u}_i)/(x_{i+1}-x_i)$ and compare the constraints. The second violated constraint C_i is chosen by finding the largest $b_{i+1}-b_i$, i, $i+1 \in \{I\}$, calculated from the $\{\hat{u}_i\}$ and so on. This method seems to be the most efficient and it is also this alternative that was used in the example above and gave the sum of squares=1778.1.

<u>Alternative 2:</u> The difference between this alternative and the preceding is the way of choosing the C_i. In the first alternative the violated constraint C_i is chosen by finding the <u>largest</u> $b_{i+1}-b_i$. In this alternative we choose C_i from the left to the right without looking at the size of $b_{i+1}-b_i$. In this case the solution of the above example will be as follows:

The data are

x_i	0	20	40	60	80	120	160	180
m_i	27	9	8	10	9	19	10	8
$\bar{y}_i$	22.94	41.58	65.46	58.81	81.74	82.15	96.59	94.01
$\frac{\bar{y}_{i+1}-\bar{y}_i}{x_{i+1}-x_i}$	0.93	1.19	-0.33	1.15	0.01	0.36	-0.13	

From this table we see that the first violated constraint which should be selected is x_i = (20,40,60). In these three points the $\bar{y}_i$-values are replaced by the weighted least squares fitted values 22.62, 43.50 and 64.38. These values are the $\{\hat{u}_i\}$, i=1, 2, 3, in the first iteration. We then calculate $(\hat{u}_{i+1} - \hat{u}_i)/(x_{i+1}-x_i)$ and compare the constraints.

x_i	0	20	40	60	80	120	160	180
$\hat{u}_i^{(1)}$	22.62	43.50	64.38	58.81	81.74	82.15	96.59	94.01
$\frac{\hat{u}_{i+1}-\hat{u}_i}{x_{i+1}-x_i}$	1.04	1.04	-0.28	1.15	0.01	0.36	-0.13	

Now the next constraint that is violated is x_i = (40,60,80). Again these three $\bar{y}_i$-values are replaced by the weighted least squares fitted values. The new values are 58.79, 67.78 and 76.77. These values are the $\{\hat{u}_i\}$, i=3, 4, 5, in the second iteration. This gives the following result:

x_i	0	20	40	60	80	120	160	180
$\hat{u}_i^{(2)}$	22.62	43.50	58.79	67.78	76.77	82.15	96.59	94.01
$\frac{\hat{u}_{i+1}-\hat{u}_i}{x_{i+1}-x_i}$	0.93	0.76	0.45	0.45	0.13	0.36	-0.13	

The next constraint that is violated is x_i = (80,120,160). Again these three $\bar{y}_i$-values are replaced by the weighted least squares fitted values. The new values are 74.24, 84.36 and 94.49. These values are the $\{\hat{u}_i\}$, i=5, 6, 7, in the third iteration. This gives the following result:

x_i	0	20	40	60	80	120	160	180
$\hat{u}_i^{(3)}$	22.62	43.50	58.79	67.78	74.24	84.36	94.49	94.01
$\frac{\hat{u}_{i+1}-\hat{u}_i}{x_{i+1}-x_i}$	0.93	0.76	0.45	0.32	0.25	0.25	-0.02	

From this table we can see that after three iterations we have obtained the concave solution. The sum of squares in this alternative is 1839.6. This result is compared to alternative 1 which gave the solution after two iterations with the sum of squares 1778.1. Therefore we conclude that alternative 1 is probably superior to alternative 2.

A simulation study using 25 000 replications was performed to examine the properties we would get for arbitrary points on the regression curve using the estimation method described in alternative 1. These results are compared to the results using the least squares estimation method.

Suppose we have a sample $y_k = \mu(x_k) + \varepsilon_k$, k=1,....,N, where $\mu(x_k)$ is an isotonic-concave regression function with equidistant x-observations in the interval $-d \le x \le 0$ and $\varepsilon_k \sim N(0, \sigma(x_k)^2)$, where $\sigma(x_k)^2$ are limited. The regression functions examined were of the type $\mu(x) = -cx^2$ for the variation interval $0.25 \le c \le 1.00$. In the simulations with the regression function $\mu(x) = -cx^2$ the normalized curvature (NC) is expressed as

$$NC = \frac{c\Delta^2}{\sigma/\sqrt{n}}$$

The simulation study was performed for

1) 15 equidistant x-observations in the interval $-7 \le x \le 0$, ie the distance between two successive x-observations was $\Delta = 0.5$.

2) $n(x_k)=1$, k=1,...,15.

3) $\varepsilon_k \sim N(0, 1)$.

BIAS:

The result from 25 000 replications shows $\hat{y}_{bias}(x_k)$, k=1,...,15, in table 10.13.

TABLE 10.13. Bias

x_i	Wu´s approximation $NC = \frac{c\Delta^2}{\sigma/\sqrt{n}}$			L-S Estimate $NC = \frac{c\Delta^2}{\sigma/\sqrt{n}}$		
	0.25	0.50	1.00	0.25	0.50	1.00
-7.0	-0.294	-0.235	-0.172	-0.284	-0.229	-0.169
-6.5	0.086	0.108	0.111	0.079	0.102	0.107
-6.0	0.109	0.094	0.063	0.106	0.093	0.062
-5.5	0.086	0.045	0.013	0.073	0.044	0.012
-5.0	0.035	0.010	-0.003	0.034	0.008	-0.003
-4.5	0.004	-0.008	-0.007	0.003	-0.008	-0.006
-4.0	-0.014	-0.016	-0.011	-0.014	-0.014	-0.010
-3.5	-0.019	-0.017	-0.011	-0.019	-0.015	-0.011
-3.0	-0.013	-0.012	-0.004	-0.012	-0.011	-0.004
-2.5	0.003	-0.010	-0.008	0.002	-0.009	-0.007
-2.0	0.030	0.003	-0.012	0.029	0.001	-0.012
-1.5	0.068	0.038	0.006	0.069	0.037	0.005
-1.0	0.105	0.089	0.058	0.101	0.090	0.057
-0.5	0.082	0.099	0.104	0.074	0.105	0.109
0.0	-0.287	-0.238	-0.172	-0.203	-0.232	-0.175

Table 10.13. Random numbers were generated so that $Y \sim N(-NCx_i^2;1)$ for 15 equidistant x-observations in the x-interval $-7.0 \le x \le 0.0$. Bias of $\hat{Y}_i$, i=1,..,15, were estimated for NC = 0.25, 0.50 and 0.75 using 25 000 replications in Wu's approximation method and concave regression.

The standard deviations of the means in the simulations lie between 0.003 - 0.006 for both the least squares solution and Wu´s approximation.

From the table we can see that bias seems to be vary slowly for both estimation methods. It is also obvious that the estimation procedures give larger bias at both ends of the interval $-7 \leq x \leq 0$ than in the middle. If we cut off a few observations in each end we get an estimate with negligable bias for the rest of the observations. Wu´s approximation gives a bias that is almost the same as the bias for LSE-method for all x-observations. From the table we can also see that the curvature doesn´t seem to influence bias except at the end points where bias gets smaller when the curvature increases.

VARIANCE:

In the same simulation studies the tendency of the variance was examined and the results are illustrated in table 10.14. From this table we see that for both estimation methods the variance of $\hat{Y}_i$ compared to σ^2 is bigger at the end points of the regression curve. The other points seem to have approximately the same variance with a slight decrease in the middle of the examined x-interval. The variance of $\hat{Y}_i$ also seem to increase when the normalized curvature increases.

If we compare the two different estimation methods, Wu´s estimation method seems to have a slightly bigger variance than the least squares estimation method for almost all x-values.

TABLE 10.14. **Variance**

x_i	Wu´s approximation $NC = \frac{c\Delta^2}{\sigma/\sqrt{n}}$ 0.25	0.50	1.00	L-S Estimate $NC = \frac{c\Delta^2}{\sigma/\sqrt{n}}$ 0.25	0.50	1.00
-7.0	0.708	0.768	0.830	0.707	0.767	0.829
-6.5	0.363	0.410	0.479	0.360	0.408	0.476
-6.0	0.268	0.322	0.404	0.268	0.320	0.401
-5.5	0.235	0.297	0.388	0.235	0.297	0.387
-5.0	0.221	0.289	0.385	0.221	0.289	0.383
-4.5	0.215	0.285	0.382	0.216	0.285	0.380
-4.0	0.212	0.284	0.387	0.214	0.284	0.384
-3.5	0.212	0.284	0.383	0.214	0.284	0.381
-3.0	0.215	0.288	0.388	0.217	0.288	0.385
-2.5	0.218	0.289	0.386	0.219	0.289	0.384
-2.0	0.222	0.291	0.386	0.224	0.291	0.384
-1.5	0.236	0.298	0.390	0.237	0.298	0.388
-1.0	0.272	0.328	0.412	0.272	0.326	0.409
-0.5	0.363	0.412	0.481	0.361	0.410	0.478
0.0	0.714	0.775	0.836	0.712	0.774	0.835

Table 10.14. Random numbers were generated so that $Y \sim N(-NCx_i^2;1)$ for 15 equidistant x-observations in the x-interval $-7.0 \le x \le 0.0$. The variances of $\hat{Y}_i$, i=1,..,15, were estimated for NC = 0.25, 0.50 and 0.75 using 25 000 replications in Wu's approximation method and concave regression.

STANDARDIZED SKEWNESS AND STANDARDIZED KURTOSIS:

In the simulation study we also observed the standardized skewness and the standardized kurtosis. Since the tendencies were the same for all values of NC we only illustrate the results for NC = 1.00 in table 10.15. From this simulation study we can see that for both estimation methods the standardized skewness, μ_3/σ^3, was not especially close to zero, ie the distributions were skew and standardized kurtosis, μ_4/σ^4, seems to be bigger near the end points than in the middle. If we compare the least squres estimation method to Wu´s approximation we see that standardized skewness and the standardized kurtosis of these two methods are almost the same. From these simulation studies we can conclude that Wu´s approximation compared to the least squares solution is much easier to use and its statistical properties are almost the same. This means that Wu´s approximation is a good alternative if we want to save computing effort.

TABLE 10.15.

	Wu´s approximation		L-S Estimate	
x_i	$\frac{\mu_3}{s^3}$	$\frac{\mu_4}{s^4}$	$\frac{\mu_3}{s^3}$	$\frac{\mu_4}{s^4}$
-7.0	-0.105	2.953	-0.105	2.954
-6.5	0.301	3.137	0.293	3.135
-6.0	0.341	3.197	0.329	3.181
-5.5	0.334	3.204	0.318	3.180
-5.0	0.328	3.178	0.309	3.159
-4.5	0.320	3.175	0.296	3.141
-4.0	0.326	3.171	0.303	3.151
-3.5	0.340	3.209	0.317	3.175
-3.0	0.332	3.117	0.311	3.096
-2.5	0.325	3.131	0.305	3.110
-2.0	0.357	3.223	0.339	3.195
-1.5	0.337	3.170	0.324	3.147
-1.0	0.363	3.230	0.352	3.218
-0.5	0.319	3.205	0.312	3.201
0.0	-0.096	2.981	-0.096	2.978

Table 10.15. Random numbers were generated so that $Y \sim N(-NCx_i^2;1)$ for 15 equidistant x-observations in the x-interval $-7.0 \le x \le 0.0$. The standardized skewness and the standardized kurtosis of $\hat{Y}_i$, i=1,...,15, were estimated for NC = 0.25, 0.50 and 0.75 using 25 000 replications in Wu's approximation method and concave regression.

10.5. Comparison between sigmoid regression and some parametric estimation methods.

In situations where the regression function is sigmoid parametric estimation methods are commonly used. In chapter 1 we have discussed a selection of such parametric regression curves. When we have an estimation problem and we intend to use a parametric function it is important to have an a priori knowledge of the regression curve being symmetric around the inflection point or not. If the assumption about this is incorrect we will get a biased estimate for each point of the regression curve. The size of this bias will differ depending on the location of the points. Therefore we will compare estimates obtained from sigmoid regression with different parametric sigmoid LS estimates.

Since often the exact shape of the curve is not known, and since the consequences can be fatal if we assume an incorrect parametric form, we have examined sigmoid regression compared with two different parametric LS estimation methods in the following situations:

1) two different symmetric sigmoid functions:
   sigmoid regression is compared to an estimated parametric regression function which is
   - i) correctly assumed to be symmetric
   - ii) incorrectly assumed to be skew,

2) a skew sigmoid function:
   sigmoid regression is compared to an estimated parametric regression function which is
   - i) incorrectly assumed to be symmetric
   - ii) correctly assumed to be skew.

10.5.1. Comparison when the regression function is symmetric.

Generally we can conclude that the choice between the logit and probit models is not easy. Some consideration has been given by Berkson (1944) and Chambers and Cox (1967) to the choice between probit and logit models with the general result that extremely large sample sizes are required to affectively discriminate between the two. In practical situations the logistic curve is probably the most used parametric function. Therefore we examined this regression function together with one that had a more deviating form than the probit model.

To examine how the estimates of each point are affected by the choice of the estimation method three simulation studies were performed. The examined regression functions were

$$\mu(x) = 10 * \frac{e^x}{1 + e^x}$$

and

$$\mu(x) = \begin{cases} 0.0 & \text{if } x \leq -2.5 \\ 2x + 5 & \text{if } -2.5 < x < 2.5 \\ 10.0 & \text{if } x \geq 2.5 \end{cases}$$

for the x-interval $-5.0 \leq x \leq 5.0$.

The notations from the corresponding simulation studies are

$\hat{\mu}_1(x)$ is an estimate obtained by a non-parametric sigmoid regression,

$\hat{\mu}_2(x)$ is the LS-estimate assuming that $\mu(x)$ is a logit function,

$\hat{\mu}_3(x)$ is the LS-estimate assuming that $\mu(x)$ is an extreme value function.

FIGURE 10.1.

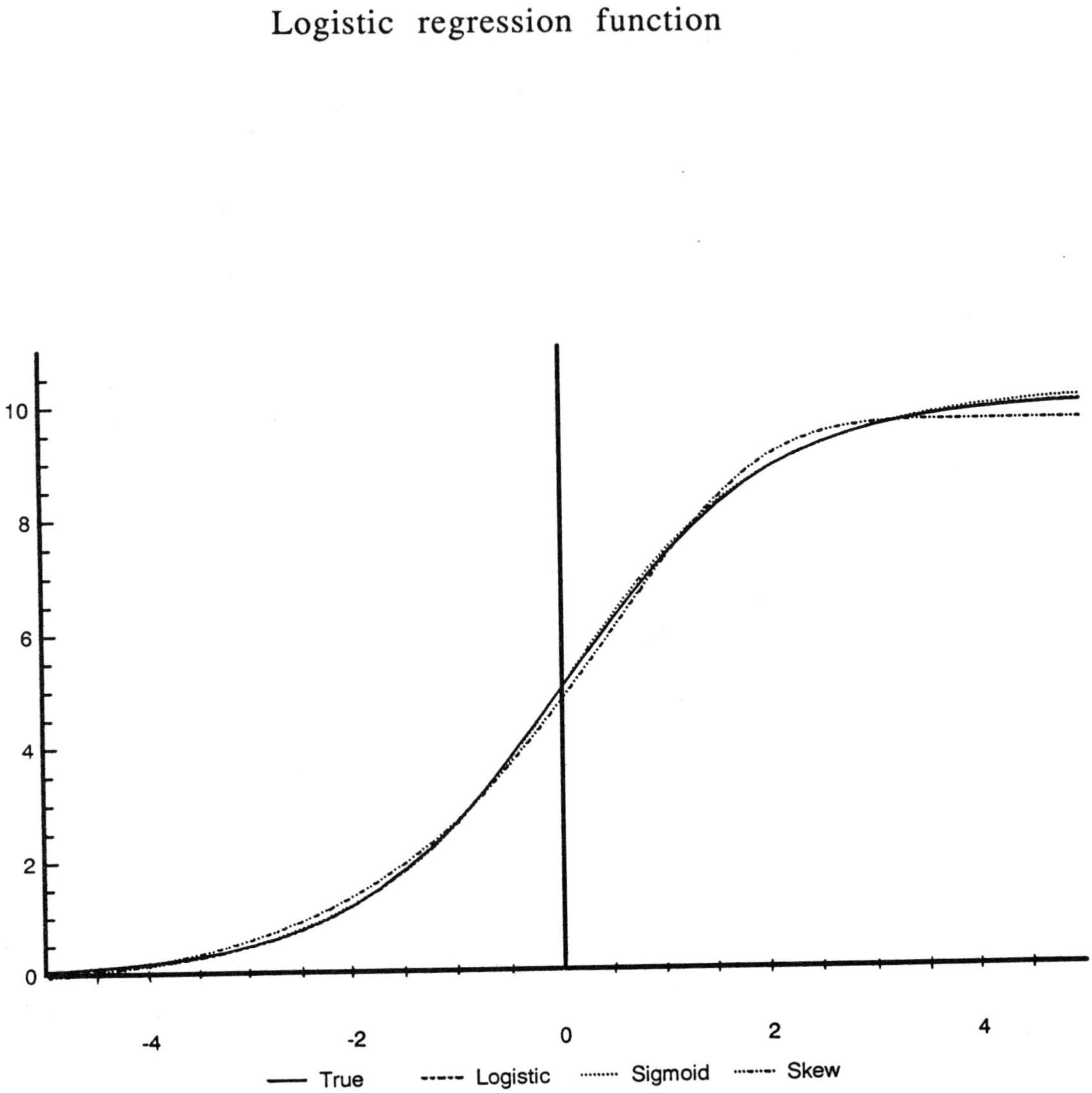

Fig 10.1. Random numbers were generated so that Y ~ N(μ(x);1), where μ(x) is a logistic regression function, for 41 equidistant x-observations in the x-interval $-5 \le x \le 5$. The distribution of $\hat{\mu}(x)$ was estimated in three different situations using 250 replications.

FIGURE 10.2.

An angular symmetric regression function

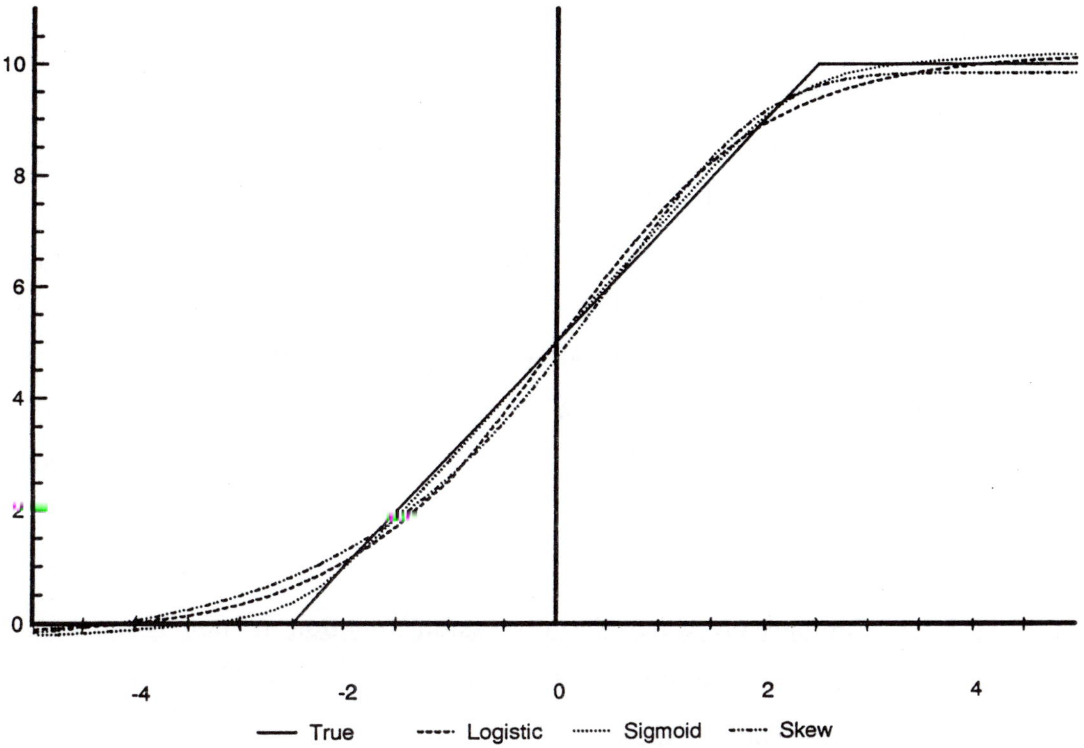

Fig 10.2. Random numbers were generated so that Y ~ N(μ(x);1), where μ(x) is an angular symmetric regression function, for 41 equidistant x-observations in the x-interval $-5 \leq x \leq 5$. The distribution of $\hat{\mu}(x)$ was estimated in three different situations using 250 replications.

In the simulations we determined the estimate of each of 41 equidistant points on the regression functions in the examined x-interval. The results from 250 replications using the first regression function are shown in figure 10.1 and from the second regression function using 250 replicates in figure 10.2. In figure 10.1 we can see that when we have a logistic function, the estimate, $\hat{\mu}_2(x)$, which assumes that $\mu(x)$ is a logit function, is the best estimate. But the estimate, $\hat{\mu}_1(x)$, obtained from our proposed estimation method, is also a very good alternative. The estimate that gives the biggest bias is the one, $\hat{\mu}_3(x)$, which assumes that $\mu(x)$ is an extreme value function. In figure 10.2 the regression function is still symmetric but deviates somewhat from the logistic curve. In this situation $\hat{\mu}_1(x)$ is the best estimate. The estimate, $\hat{\mu}_2(x)$, which assumes a logistic form is also rather good while $\hat{\mu}_3(x)$ still gives the biggest bias. From these two simulation studies we can conclude that if we don´t know the shape of the regression curve for sure our proposed estimation method is the best alternative.

10.5.2. Comparison when the regression function is non-symmetric.

Of course we are also interested in what bias we would get if the regression function is skew.

To be able to compare the results from the simulation studies using a symmetric regression function with those obtained from a non-symmetric situation three new simulation studies were performed. The examined non-symmetric regression function was $\mu(x) = 10*[1-\exp(-\exp(x))]$ for the x-interval $-5.0 \le x \le 5.0$.

The notations from the corresponding simulation studies are analogous to those in the preceding section namely

FIGURE 10.3.

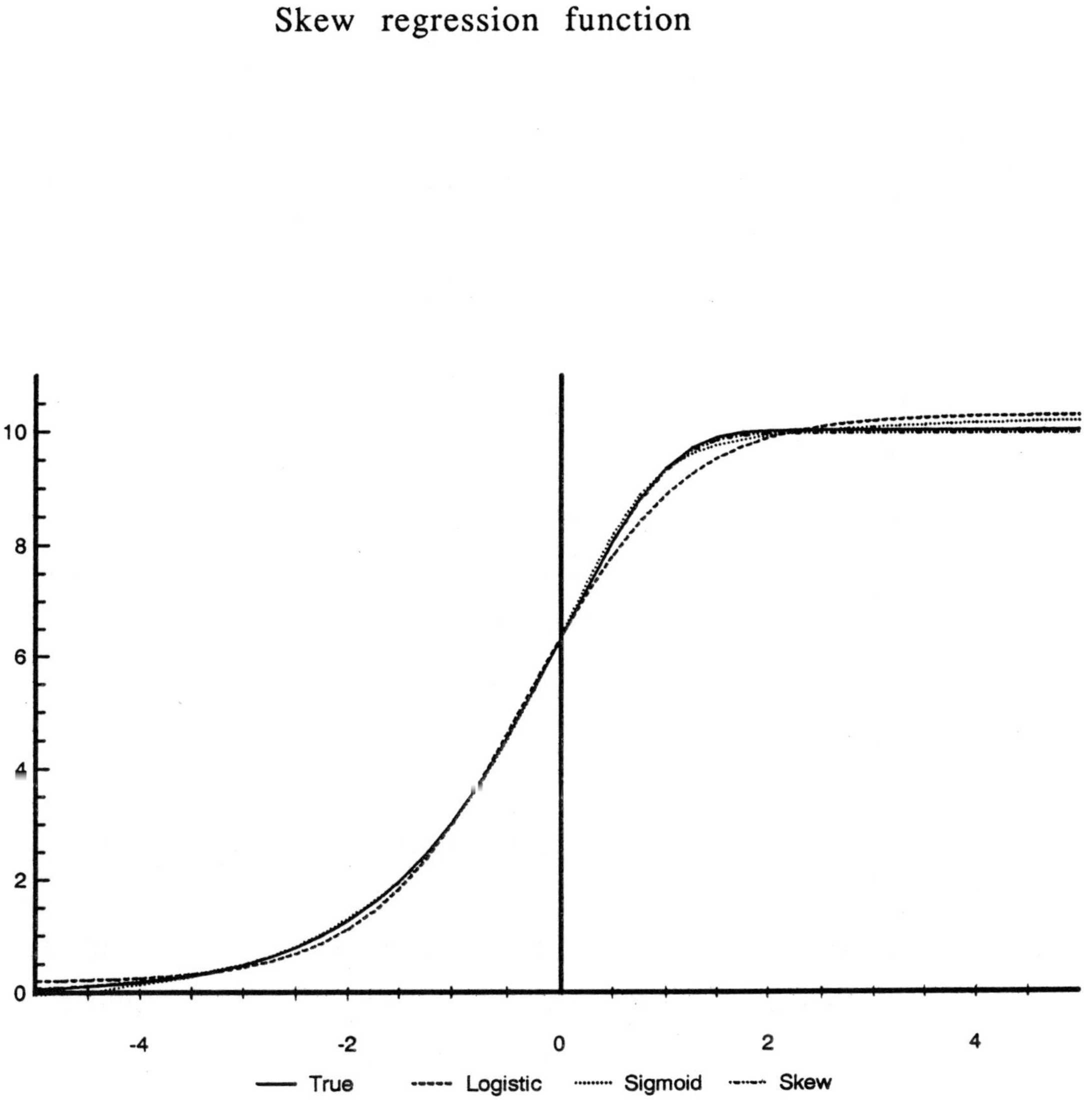

Fig 10.3. Random numbers were generated so that Y ~ N(μ(x);1), where μ(x) is a skew regression function, for 41 equidistant x-observations in the x-interval -5 ≤ x ≤ 5. The distribution of $\hat{\mu}(x)$ was estimated in three different situations using 250 replications.

$\hat{\mu}_1(x)$ is an estimate obtained by a non-parametric sigmoid regression,
$\hat{\mu}_2(x)$ is the LS-estimate assuming that $\mu(x)$ is a logit function,
$\hat{\mu}_3(x)$ is the LS-estimate assuming that $\mu(x)$ is an extreme value function.

In the simulations we determined the estimate for each of 41 equidistant points on the regression functions. The results from 250 replications are shown in figure 10.3. In this figure we have used a skew regression function. This resulted in a situation where $\hat{\mu}_3(x)$, assuming a skew curve, was the best estimate. The estimate, $\hat{\mu}_1(x)$, obtained from our proposed estimation method, was the second best choice. The estimate, $\hat{\mu}_2(x)$, assuming a logit model, gave the biggest bias. Comparing these results with the results in the preceding simulation studies we can conclude that in practical situation, where the regression function is unknown, our proposed estimation method is the better choice.

10.6. Comparison between sigmoid regression, logistic regression and a discrete alternative to the logistic regression proposed by Nash.

Of course we are interested in comparing sigmoid regression with a few other sigmoid estimation methods in some practical situations. In Nash (1977) two examples are given where logistic regression is compared to a discrete alternative. In this section we will compare the results from this article and show the corresponding results using sigmoid regression. But first we will explain the principles of the discrete estimation procedure proposed by Nash.

10.6.1. The estimation method proposed by Nash.

The logistic growth function can be written as

$$v(t) = \frac{h}{1 + \exp(p+qt)}$$

where t is the independent variable and h, p and q are parameters. This function can be obtained from the integration of the logistic differential equation

$$\frac{dv}{dt} = -qv + \frac{qv^2}{h} = \frac{-qv}{h}(h-v)$$

By forming the difference equation corresponding to this equation it is possible to derive a discrete alternative, u(t), which at the points where it is defined, has values similar to the logistic.

$$u(t+1) = (1-q)\ u(t) + \frac{q(u(t))^2}{h}$$

The parameter p is now only embedded in

$$u(0) = v(0) = \frac{h}{1 + \exp(p)}.$$

This value must of course be computed first. The estimation of the parameters q and h is done by a full iterative non-linear LS method, which agrees with the estimation method of the conventional logistic described in Oliver (1964):

Suppose we have a series $\mu(t)$, t=1,....,N, between the bounds g and g+h. Then the sum of squares

$$Q = \sum_{t=1}^{N} [u(t) - \mu(t) + g]^2$$

can be minimized providing that the partial derivatives of u(t) with respect to the parameters can be computed. That this can be done is shown in Nash (1977).

10.6.2. Two examples using the discrete alternative estimation method proposed by Nash.

In his presentation of the estimation method Nash demonstrates the procedure by giving the solutions to two examples. He also uses these examples to compare the discrete alternative with the logistic estimation.

EXAMPLE 1.

A sample of 23 annual compilations of production of milk per cow (in lb) was taken from the Canadian dairy herd. The bounds were determined to g = 3950 and g+h = 15000. The data points were

6111, 6220, 6141, 6086, 6007, 6080, 6074, 6217, 6500, 6743, 6594, 6734, 6759, 6935, 6993, 6935, 7218, 7400, 7553, 7851, 7854, 7738, 7930.

The logistic function, that was fitted, had q = -0.0452647, p = 1.61689 and v(0) = 1830.26. This gives the sum of squares Q = 521739.

The discrete alternative function had q = -0.0457614 and u(0) = 1831.67 with the sum of squares Q = 519507.

Sigmoid regression. When we used sigmoid regression we obtained the following result:

Regression coefficients:	Bending points:
0.0	1
1.3	5
96.5	6
143.7	16
167.5	17
267.3	19
12.1	20
0.0	23

This solution gives the sum of squares Q = 159800.

If we compare the three estimation methods we can see that sigmoid regression gives the smallest sum of squares. This depends partly on the fact that the estimates obtained from sigmoid regression are not symmetric around the estimated inflection point.

EXAMPLE 2.

In this example Nash gives 12 data points coming from a situation where a continuous model was expected to apply.

5.308, 7.240, 9.638, 12.866, 17.069, 23.192, 31.443, 38.558, 50.156, 62.948, 75.995, 91.972.

The logistic function obtained q = -0.31357, exp(p) = 49.0916, h = 196.186 and v(0) = 3.91654. This gives the sum of squares Q = 2.58726.

The discrete alternative function had q = -0.361421, h = 165.584 and u(0) = 3.97726 with the sum of squares Q = 2.79904.

Sigmoid regression. When we used sigmoid regression we obtained the following result:

Regression coefficients:	Bending points:
1.93	1
2.40	2
3.23	3
4.20	4
6.31	5
7.68	6
11.41	8
12.79	9
13.05	10
15.98	11

This solution gives the sum of squares Q = 0.2151.

If we compare the sum of squares obtained from the three estimation methods we can see that sigmoid regession gives by far the smallest sum of squares in this example too. Therefore we conclude that sigmoid regression is superior to other estimation methods used today in those applications.

REFERENCES.

BARLOW, R.E., BARTHOLOMEW, D.J., BREMNER, J.M. and BRUNK, H.D.; (1972): Statistical Inference Under Order Restrictions, Wiley, New York.

BERKSON, J.; (1944): Application of the Logistic Function to Bio-Assay, Journal of the American Statistical Association 39, pg 357-365.

BICKEL, P. and DOKSUM, K.; (1977): Mathematical Statistics, Holden-day.

CHAMBERS, E.A. and COX, D.R.; (1967): Discrimination between Alternative Binary Response Models, Biometrika 54, pg 573-578.

DAHLBOM, U.; (1983): Consistency of Unimodal and Monotonic Regression, Seminar paper 1983:6, University of Göteborg Department of Statistics. (in Swedish).

DAHLBOM, U.; (1986): Some Properties of Estimates of Unimodal Regression, Research Report 1986:6, University of Göteborg, Department of Statistics.

DENT, W.; (1973): A Note on Least Squares Fitting of Functions Constrained to be Either Nonnegative, Nondecreasing or Convex, Management Science 20, pg 130-132.

FINNEY, D.J.; (1978): Statistical Methods in Biological Assay, Charles Griffin & Co, London.

FRASER, D. A. S. and MASSAM, H.; (1987): An Algorithm for Concave Regression, Foundations of Statistical Inference, D. Reidel Publishing Company.

FRASER, D. A. S. and MASSAM, H.; (1989): A Mixed Primal-Dual Bases Algorithm for Regression Under Inequality Constraints. Application to Concave Regression, Scandinavian Journal of Statisticts: pg 65-74.

FRISÉN, M.; (1980): U-Shaped Regression, COMPSTAT Proc Comput Statist 4: pg 304-307.

FRISÉN, M.; (1986): Unimodal Regression, The statistician, Vol. 35: pg 499-505.

GOMPERTZ, B.; (1825): On the Nature of the Function Expressive of the Law of Human Mortality and on a New Method of Determining the Value of Life Contingencies, Philosophical Transactions of the Royal Society, vol XXXVI pg 513-585.

HANSON, D.L. and PLEDGER, G.; (1976): Consistency in Concave Regression, Annals of Statistics 4, pg 1038-1050.

HILDRETH, C.; (1954): Point Estimates of Ordinates of Concave Functions, Journal of American Statistical Association 49, pg 598-619.

HOLLOWAY, C.A.; (1979): On the Estimation of Convex Functions, Operations Research 27, pg 401-407.

HOLM, S. and FRISÉN, M.; (1985): Nonparametric Regression with simple curve characteristics, Research report 1985:4, University of Göteborg, Dept. of Statistics.

MAMMEN, E.; (1988): Nonparametric Regression under Qualitative Smoothness Assumptions, Unpublished Manuscript, Universität Heidelberg, Institut für Angewandte Mathematik.

McCULLAGH, P. and NELDER, J.A.; (1983): General Linear Models, Chapman and Hall, New York.

NASH, J.C.; (1977): A Discrete Alternative to the Logistic Growth Function, Applied Statistics 26, pg 9-14.

OLIVER, F.R.; (1964): Methods of estimating the Logistic Growth Function, Applied Statistics 13, pg 57-66.

ROBERTSON, T., WRIGHT, F.T. and DYKSTRA, R.L.; (1989): Order Restricted Statistical Inference, Wiley, New York.

SCHMOYER, R.; (1984): Sigmoidally Constrained Maximum Likelihood Estimation in Quantal Bioassay, JASA 1984 (79), pg 448-453.

STONE, R.; (1979): Sigmoids, Bullentin in Applied Statistics, 1980 (7), pg 59-119.

VERHULST, P.F.; (1838): Notice sur la Loi que la Population Suit dans son Accroissement, Correspondance Mathématique et Physique Publieé par A. Quetelet, tome X, pg 113-121.

WINSTON W.L.; (1991): Operation Research Applications and Algorithms, PWS-Kent Publishing Co, Boston.

WU, C-F.; (1982): Some Algorithms for Concave and Isotonic Regression, Studies in the Management Sciences 19: pg 105-116.